MW00675307

D.E. MOGGRIDGE

BRITISH INTERNATIONAL GOLD MOVEMENTS AND BANKING POLICY, 1881-1913

BY

W. EDWARDS BEACH

ASSISTANT PROFESSOR OF ECONOMICS IN WILLIAMS COLLEGE

GREENWOOD PRESS, PUBLISHERS
WESTPORT, CONNECTICUT

The Library of Congress has catalogued this publication as follows:

Library of Congress Cataloging in Publication Data

Beach, Walter Edwards.
   British international gold movements and banking
policy, 1881-1913.

   Original ed. issued as vol. 48 of Harvard economic
studies series.
   Bibliography: p.
   1. Currency question--Gt. Brit. 2. Gold.
3. Banks and banking--Gt. Brit. 4. Business cycles.
I. Title. II. Series: Harvard economic studies,
v. 48.
HG939.B352 1972   332.4'1'0942   76-138201
ISBN 0-8371-5554-1

To

MY MOTHER AND FATHER

PREFACE

THE problems connected with international gold movements have received renewed attention as a result of the breakdown of the gold standard during the depression following 1929. In a world where friction is so important in the economic sphere, and where international capital is very mobile, an automatic standard is not attainable. It has not been recognized clearly enough, however, that the pre-war standard was not strictly automatic. The problems have grown quantitatively during the post-war years, but most of them were faced by the Bank of England in the pre-war era, especially after 1900. The growth of foreign balances in London, the rise of other nations as competitors of England in the export markets for manufactured goods, and the increasing friction caused by greater rigidity in the price and wage structure and the growth of high tariffs presented problems of management to the Bank of England similar to those encountered today.

Cyclical movements of gold coin from New York to the interior of the country, accompanied by imports of gold from abroad during periods of business prosperity, were encountered in the United States during the pre-war era. In depression both flows were reversed. The importance of these phenomena for problems of money and banking and international trade was pointed out by the late Professor Allyn A. Young. At his suggestion, the present study was begun to see whether the same relations between internal and external movements of gold were to be found for England.

During the period selected for study there were five clearly marked complete business cycles, and the movements of gold and currency are considered in relation to business conditions. By the beginning of the period the Bank of England had mastered the principles of action through which it endeavored to protect its reserve and control the money market. Except for the South African conflict, the period was free from war.

I am deeply indebted to Professor Young for his advice and criticism during the preparation of this study. After his death the work was continued under the direction of Professor John H. Williams, and I wish to express my appreciation of the many helpful suggestions which he has made. I also wish to thank Professor S. E. Harris of Harvard University and Professor James W. Angell of Columbia University for their criticism of the manuscript.

W. E. B.

WILLIAMSTOWN, MASSACHUSETTS
    December, 1934

CONTENTS

CHAPTER I

CHAPTER II

Ricardo, 7. — The classical theory of gold movements, 8. — Professor Taussig and the neo-classical position, 9. — The neo-classical position related to cyclical fluctuations of business, 10. — Hawtrey's application of the classical theory to cyclical conditions, 12. — Influence of foreign exchange writers upon the theory of gold movements; Goschen and Laughlin, 19. — Angell's theory of the transfer of purchasing power, 26. — The application of the Ricardian theory of gold flows by the Currency School, 31. — Contributions of the Banking School, 32. — Influence of the Currency School upon the treatment of cyclical fluctuations in gold flows by de Laveleye and Juglar, 35. — Influence of the rate of discount upon prices and specie movements, 36.

CHAPTER III

Cyclical fluctuations in business, 39. — Statistics of foreign gold movements, 43. — Internal movements of currency, 50. — Accuracy of statistics, 64. — Cyclical fluctuations in gold and currency, 74.

CHAPTER IV

Reserves of joint stock and private banks, 80. — Cash and balances at the Bank of England, 82. — Accumulation of gold by joint stock banks after 1900, 87. — Bank deposits, 94. — Interest rates, 100.

CHAPTER V

Description of business cycles of 1881–1885, 105. — Gold and currency movements, 111. — Effects of Bank policy upon the money market, 113. — The revival of business from 1886 to 1890, 113. — Gold and currency movements, 114. — The Baring crisis, 115.

CHAPTER VI

CHAPTER VII

CHAPTER VIII

CHAPTER IX

LIST OF TABLES

LIST OF CHARTS

BRITISH INTERNATIONAL GOLD MOVEMENTS
AND BANKING POLICY, 1881–1913

CHAPTER I

INTRODUCTION

International gold movements are important in periods of disordered currencies and dislocated exchanges, but the transfers are likely to be spasmodic and they play little part in the mechanism of trade adjustment. So long as gold imports and exports may be made freely by banks, the gold standard acts as a regulator of prices and exchanges, whether or not gold coin is used, and whether or not other forms of currency are convertible into gold. Under certain conditions the use of gold coin and conversion of notes to gold or *vice versa* may impede, rather than aid, the working of the gold standard. The early classical explanation of the problem of trade adjustment was based upon the premises of a relatively simple economic structure, where credit and other forms of currency acted in a manner similar to what would have been expected of a pure metallic money, and a price-cost structure which was extremely sensitive to alterations in the volume of currency and credit. The price-specie flow mechanism also implied the predominence of commodity trade in the international accounts of nations, and adjustment of disequilibrium was assumed to occur almost entirely through the effects of gold shipments upon the price level and ultimately upon the volume of international trade in goods. Adjustments through forces operating within the gold points were not regarded as important.

While the views of the earlier classicists have been modified by later writers, the price-specie flow mechanism is still the core of the doctrine of adjustment of disequilibria. The importance of short-term capital movements and speculative operations in exchange are recognized, but gold flows are regarded as the important element in the restoration of equilibrium when there has been a force of considerable quantitative importance at work to upset the original condition of equilibrium. It is also recognized that under modern credit banking the response of prices to specie movements may be less sensitive than in a primitive state where standard metal alone is used in internal circulation, since various offsetting operations by central banks or lags in the response of bank credit and prices may occur unless the banking systems are "loaned up." The narrowing of the gold points as a result of the development of international banking connections and improvement in transportation has served to increase the ease with which specie is moved, and less adjustment is possible within the gold points. The neo-classical position is strengthened as a consequence of this development. On the other hand, the concomitant growth of short-term borrowing between financial centers, and the expansion of speculation in exchange, in "international" securities and in commodities with a "world market" have reduced the volume of specie movements, and enabled greater adjustment to occur within the narrowed gold points. The question of the price-specie flow analysis may be attacked from an historical standpoint, and its validity determined upon the basis of the institutional developments which have occurred. Or the problem may be attacked from a logical viewpoint, and a different explanation of the process of adjustment worked out irrespective of the development of institutions.

The second method is illustrated by the work of writers [1] who stress the shifting of demand schedules, rather than the mechanism of sectional price levels, as the means by which adjustment of disequilibrium is achieved. Writers on foreign exchange have also pointed the way to a different explanation by their emphasis upon discount rates as the major force behind the movement of specie. Angell's theory [2] of the transfer of purchasing power via the bill market is a logical development from the position of foreign exchange specialists, although it depends in part upon the development of close international banking connections and the possibility of adjustments within the gold points.

The classical price-specie flow mechanism was designed to explain conditions which economists describe as "long-run," and

[1] See, e.g., Harry D. White, *The French International Accounts, 1880–1913*, Cambridge, Massachusetts, 1933, and Bertil Ohlin, *Interregional and International Trade*, Cambridge, Massachusetts, 1933.

[2] James W. Angell, *The Theory of International Prices*, Cambridge, Massachusetts, 1926.

was not applied directly to the phenomenon of business cycles by the earlier writers. But adaptation of the classical doctrine to cyclical fluctuations began during the nineteenth century, and other writers have pointed out inconsistencies in the application of the price-specie flow analysis to cyclical phenomena. Many new questions are raised when gold movements are considered as a problem in connection with business cycles, and a considerable portion of the present study is concerned with the relations between specie flows and the volume of domestic credit and currency.

Attention was frequently called by Professor Allyn A. Young to a peculiar movement of gold in the United States under the National Banking System.[1] New York City was an intermediary through which demands for the rest of the country for money were transmitted to the international market. Gold imports coincided with drains of cash to the interior, and a reflux of money to New York and exports to foreign countries were concurrent phenomena. There was also a close correlation between these flows and movements of bank loans and deposits in New York and the rest of the country. A definite inverse relationship in the various items was apparent when cyclical fluctuations in New York were compared with fluctuations for the rest of the United States. And the relation of New York to the international loan market changed during the progress of the cycle.

The flow of gold during cyclical fluctuations also appeared to be at variance with what was to be expected according to the classical analysis of long-time movements. Professor Young found that during periods of prosperity gold tended to come into the United States in response to high discount rates, even though a relatively high price level should have caused gold to be exported. The question of the relative importance of discount rates as compared to commodity prices in bringing about the movement of specie is raised so far as cyclical fluctuations are concerned, and the present study attempts to evaluate the quantitative problems involved. It has been suggested that possibly creditor nations

[1] *An Analysis of Bank Statistics for the United States*, Cambridge, Massachusetts, 1928, esp. p. 28.

may differ from debtor countries in the manner in which specie movements occur over the course of business cycles,[1] and the study of the British pre-war data may be useful in answering this query.

In addition to questions of gold movements, attention has been paid to the problem of banking arrangements and their effect upon the control exercised by the Bank of England. The development of branch banking exerted a profound change in the relations of the central bank to the joint stock banks and to the money market, and enabled the Bank of England to control the market adequately in general, despite growing difficulties arising from the increasing importance of other nations as trade rivals to Great Britain and the strain introduced by the growth of foreign balances in London. The slender reserves with which the Bank operated prior to 1892 would have been inadequate if foreign funds had been as large as in later periods. Central banking problems have been increased greatly by the growth of such balances, and the transfer of these funds from one market to another have been one of the chief causes for the failure of the gold standard to work in an automatic manner. It is only gradually that it is being recognized that the pre-war gold standard was not automatic in a rigid sense, and that the post-war problems are different not in kind but in degree from those encountered in the period prior to 1913. It is hoped that the present study may throw light on the working of the gold standard in this earlier period.

In Chapter II a brief summary is made of the more important doctrines with respect to the role of gold movements as the agent in bringing adjustment of disequilibria in the international balance of payments.[2] In the succeeding portions of the book the external and internal movements of gold are examined, followed by a survey of the London money market and the conditions of credit during the period 1881–1913. Finally an attempt is made to set forth a theory of gold movements over cyclical periods.

---

[1] See, e.g., the remarks of J. M. Keynes, *The International Gold Problem*, London, 1931, p. 188.

[2] The definition of "balance of payments" presents difficulties. I have used the term to refer to immediate obligations only, and have used the term "international accounts" when referring to the total of all items on the balance sheet of a nation. See White, *op. cit.*, p. 38.

CHAPTER II

THE THEORY OF INTERNATIONAL GOLD MOVEMENTS

(1) A brief survey of some of the more important ideas concerning the part played by specie flows in the adjustment of international trade disequilibria is made in the present chapter.[1] The classical and neo-classical theories of Mill and Taussig are considered, followed by an examination of the expansion of the classical doctrine to problems of the business cycle made by Hawtrey and other writers.

Before turning to the classical theory of trade adjustment, one point suggested by Ricardo will be considered. Ricardo maintained that interest rates do not depend upon the level of prices,[2] but changes in the volume of money *would* affect interest rates temporarily.[3] Clearly he means here market rates of discount, which would affect both long- and short-term security prices for the moment, and there is some implication of a distinction between the price levels of capital goods and liquid capital and of consumable goods. While Ricardo does not examine the question further, the notion is implicit that a redundancy of currency [4]

[1] The survey of the literature made by Angell, *op. cit.*, has been indispensable as a guide.

[2] David Ricardo, *The Principles of Political Economy and Taxation*, London, 1817, Everyman's edition, London, 1912, pp. 84–86, and *The High Price of Bullion*, reprinted in *Economic Essays*, London, 1923, edited by E. C. K. Gonner, pp. 32–33.

[3] Ricardo, *The High Price of Bullion*, p. 35.

[4] By redundancy Ricardo means that prices in one country are high relative to those of other countries. International trade adjustment is treated in terms of redundancy, and the unilateral theory of gold flows held by Ricardo (see Angell, *op. cit.*, pp. 56–58, 64, and 70; Jacob Viner, *Canada's Balance of International Indebtedness, 1900–1913*, Cambridge, Massachusetts, 1924, pp. 193 ff.; and David Ricardo, *The High Price of Bullion, op. cit.*, pp. 4–7, 11, and 12) is illustrative of the emphasis placed upon the monetary aspects of the problem. Angell has pointed out contradictions in Ricardo's position (*op. cit.*, p. 70), and suggests that the inconsistency in the treatment of foreign exchange premiums and the unilateral specie flow doctrine may be reconciled "by regarding that state of relative prices which brought about the deficit in the balance of payments as being itself symptomatic of a relative redundancy or deficiency in the currencies" (*loc. cit.*). This view is strengthened by the following passage from a letter of Ricardo to Malthus: "This relative redundancy may be produced as well by a diminution of goods as by an actual increase of

used for purchasing consumable commodities does not necessarily imply a redundancy of currency for other purposes, particularly for short-term funds. The converse situation might also occur where short-term interest rates were relatively high in one country, even though there was no redundancy of currency used for the purchase of consumable goods.

(2) The classical bi-lateral theory of gold movements became the accepted doctrine early in the nineteenth century. It is finally found in a complete and clear form in the work of Mill, although all of the essential elements are seen earlier. Gold flows may be either the cause or the result of an unfavorable balance of trade.[1] Any important disequilibrium in the balance of trade will cause gold to flow, and the movement of specie will set in motion forces which remove the initial cause of the disturbance. Mill also held that an annihilation of credit equivalent to the subtraction of coin or notes from circulation is equally effective in altering the level of prices and correcting the disequilibrium in the balance of trade.[2] A slight disparity between the demand for and the supply of bills of exchange may not lead to gold flows, since the premium on exchange may bring a sufficient corrective force operating within the gold points to remedy the disequilibrium in the balance of trade.

The price-specie flow analysis depends upon the quick response of prices and costs of production to movements of specie, and a prompt alteration in the volume of international trade to correct the initial disturbance of the equilibrium in the balance of payments. Additional elements have been added by later writers,

money (or what is the same thing, by an increased economy in the use of it) in one country; or by an increased quantity of goods or by a diminished amount of money in another." *Letters of David Ricardo to Thomas Robert Malthus*, Oxford, 1887, edited by James Bonar, p. 11. See also the letter, pp. 20-23, which is referred to by Angell (*op. cit.*, p. 58). Apparently a diminution of goods creates a redundancy of currency, which leads to a flow of specie, which in turn creates the unfavorable balance of payments. This chain of events is consistent with Ricardo's treatment of foreign exchange premiums, but it does not dispose of his denial that a crop failure could create a redundancy of currency (see Viner, *op. cit.*, p. 193).

[1] J. S. Mill, *Principles of Political Economy*, London, 1848; Ashley edition, London, 1920, pp. 620-621 and 623-624.

[2] *Ibid.*, pp. 617-618. See also Ch. XII, esp. pp. 524-525.

but the neo-classical school relies upon the movement of specie to set in motion the forces which restore equilibrium whenever the departure from equilibrium is quantitatively important.

(3) The neo-classical position set forth by Taussig[1] makes allowance for the sensitivity of banking systems. When the banking systems are "loaned up," the volume of bank deposits and notes is sensitive to changes in reserves, and gold movements will dominate the level of prices.[2] But when the relation between reserves and the media of payments is elastic, specie flows may not exert an immediate influence upon commodity prices, and the effect of the flows will be dependent to a considerable degree upon the temper of the business community.[3] Discount rates may be the proximate cause of specie movements over short periods, and gold may be drawn into a country on account of the depletion of bank reserves.[4] Such flows may be eliminated to a considerable extent by the transfer of bankers' balances, or the exchange of international securities.[5] The underlying forces, however, are concerned with relative price levels, and the ultimate adjustment of any important departure from equilibrium comes through price changes and alterations in the trade balance.

The effects of real changes in the positions of lending and borrowing countries are considered by Taussig,[6] and he points out that in some cases an agreement, tacit or otherwise, is made by which the borrower employs the funds lent in purchasing goods in the lending country.[7] The flow of goods proceeds immediately, and the relative price levels of the two countries need not change in order to induce commodity movements. Ordinarily, however, it is assumed that new foreign loans lead to specie flows, and consequent changes in relative prices, which alter commodity movements.[8] Where the connection between bank reserves and deposits is loose, the chronological order of events in connection with

[1] F. W. Taussig, *International Trade*, New York, 1928. See also White, *op. cit.*, Ch. I.
[2] Taussig, *op. cit.*, pp. 198–207.
[3] *Ibid.*, p. 202.
[4] *Ibid.*, pp. 207–208.
[5] *Ibid.*, Ch. 18.
[6] *Ibid.*, Chs. 11 and 12.
[7] *Ibid.*, pp. 124–125.
[8] *Ibid.*, pp. 123–130.

specie flows may be different from that assumed in the classical statement: an influx of specie may follow, rather than precede, an increase in the volume of the media of payments and a rise of prices.[1] This notion is applied specifically to the process of transferring an international loan, and is linked up with cyclical conditions. "If, for example," Taussig states,

the country is in the early stage of large borrowings — if a succession of loans from abroad is in course of being contracted — the proceeds of the loans will provide the means for paying for added imports. While advancing prices will tempt imports, the loans, so long as they continue, enable them to be paid for from year to year. They may even enable still more specie to flow in, and so give support to a still further advance in prices and still further imports. The process may go on *crescendo*, until at last there is a halt in the lending operations, not unlikely to be sudden, and marked by a financial and industrial crash.[2]

In the case of Great Britain the greatest volume of capital exports took place when the movement of prices at home was upward, and ". . . the export of goods apparently has taken place, not in connection with a cheapening of goods in the lending country, but in spite of the fact that its goods have seemed to be dearer at times of great capital export." [3] Whether large capital exports were associated with *relative* increases in export *versus* import prices or not, the fact that they occurred principally during periods of rising prices in the lending countries is significant for the development of a theory of gold movements concerned with cyclical fluctuations.

Taussig's analysis provides such a theory, with capital movements dominating the situation. Borrowing nations would receive gold during the expansion phase of a business cycle, despite a high level of commodity prices relative to lending countries. This borrowing would enable expansion to continue until the loans ceased, generally because of a crisis. The cessation of loans would not necessarily stop the influx of gold, since the high discount rates, which are the proximate cause of the gold movements, will continue until the crisis is over.

---

[1] *Ibid.*, pp. 207–209.   [2] *Ibid.*, p. 209.
[3] *Ibid.*, p. 239. See also Ch. 21. See, however, A. G. Silverman, "Some International Trade Factors for Great Britain, 1880–1913," *Review of Economic Statistics*, August, 1931.

The magnitude and duration of the flow of loans will be affected by changes in the sectional price levels, in addition to other forces. The barter terms of trade may become either more or less favorable to the lender. If import prices rise relatively to export prices (less favorable terms of trade), the transfer of the loan to the borrower is made with a minimum of friction.[1] The flow of specie necessary to maintain equilibrium in the balance of payments will be small relative to the magnitude of the loans. To the extent that the drain on the specie reserves of the lender is the limiting factor to the flow of loans, the duration and magnitude of the loan operations will be greater under the conditions above than in the case where a larger flow of specie is necessary to maintain equilibrium in the balance of payments. If the prices of import goods decline relatively to prices of export goods in the lending country (more favorable terms of trade), the flow of specie must be larger relative to the loans, since adjustment of the trade balance is impeded by an increase of exports by the borrower.[2] The duration of the loan movement will be shorter than in the first case, since the reserves of the lender will be drawn down more rapidly. In both cases, however, the import of specie by the borrower is associated with a greater expansion of production and a higher price level than are found in the lending nation.

In depression the borrowing country would lose gold, even though its price level was low relative to that of creditor nations. The proximate cause of the specie flow would be a relative decline of discount rates, but the underlying force would be the pressure upon the balance of payments created by the interest and amortization payments on old loans in the absence of new borrowing. Loss of specie is consistent with a price level below that of lending countries.

The theory set forth briefly here assumes that the response of prices and costs is not immediate following changes in discount or interest rates. If the price-cost structure is sensitive to fluctuations in money rates, the expansion of business and the rise of

---

[1] It is assumed that the elasticity of demand for international goods is approximately the same in the two countries.

[2] A shift of demand schedules for international goods might make adjustment easier.

prices in borrowing countries during the recovery and prosperity stages of the cycle could not continue long, and in depression the adjustment downward of prices would enable the borrower to transfer interest payments without large gold exports.[1]

The doctrine set forth here may also be related to the internal movements of currency and credit. In the borrowing nation, a flow of funds from the central money market to the interior would be expected during the period of expansion. At the same time funds would be flowing in from abroad, and specie imports would be coupled with an expansion of deposits and currency in the interior of the country.[2] The two types of flow were recognized early in the nineteenth century. The two movements were sometimes regarded as concurrent, however, the drain of gold abroad occurring at the same time that notes and coin were moving to the interior of the country. The analysis of the question made by R. G. Hawtrey[3] suggests that this is the phenomenon to be expected during the expansion phase of a business cycle; but before examining this idea it is necessary to consider his general theory of money and credit.

(4) Prices are determined, in Hawtrey's view, by the expenditure of purchasing power by consumers out of their income. More specifically, price changes vary in accordance with the amount of the "unspent margin", which is the difference between the income of consumers and their outlay, plus the balances at the beginning of the period of consumers and business men. This margin can be equated to the amount of money in circulation, plus the volume of bank deposits outstanding. During a period of expanding credit he finds that the consumers' incomes rise with the increase in business. A gradual drain of money into circulation results, to be succeeded eventually by higher discount rates, and ultimately by contraction as consumers' incomes and outlays fall off.[4] Except for the introduction of money there would be no

---

[1] Cf. below, pp. 13-14.
[2] See above, pp. 5-6.
[3] *Currency and Credit*, London, 1919. I have used here the third edition (London, 1928).
[4] *Ibid.*, Ch. IV.

end to an expansion, for credit is inherently unstable. The function of the former is to supply a means of discharging a debt legally, and to stabilize the unit of value.[1]

He goes on to show that in a country depending upon gold for its supply of money, the credit structure might break down because of the exhaustion of its supply of legal tender. He summarizes this concept in the following words:

> The grounds for this conclusion were two fold. First, though the consumers' income is increased, the part composed of wages is not immediately increased in proportion, and even when wages do rise, the amount of cash continuing in circulation, the cash portion of the unspent margin, only grows gradually. There is thus a long interval between an increase in the quantity of credit and the corresponding increase in the circulation of money. Secondly, the increase in the consumers' outlay tends to deplete stocks of goods, orders are given for the replenishment of stocks, there supervenes an over-pressure upon the productive powers of the country and an accumulation of orders awaiting execution, and this accumulation of arrears of orders constitutes a latent demand for credit, which will only materialise when the arrears begin to be overtaken.[2]

Thus a credit expansion is dangerous, not only because it reduces reserves of legal tender at the moment as notes and coin are drawn into circulation, but because there are *latent* demands for credit and cash which will develop even after the contraction of new orders has diminished.

Hawtrey then turns to the international aspects of the question. As a country experiences a credit expansion its trade balance tends to become unfavorable. As a result, through a mechanism to be described later, gold flows abroad; thus to the inward drain of cash there is added an efflux abroad as well. But this gives bankers prompt warning, and they raise discount rates and curtail credit, and so prevent an undue expansion.[3]

But there is nothing to prevent a world-wide expansion of credit. For in that case no nation would lose gold, except momentarily. If all nations are passing through the same period of expansion of credit during the prosperity stage of a business cycle, the flow of gold means merely that they are not keeping precisely

---

[1] *Ibid.*, Chs. I and II.
[2] *Ibid.*, p. 107. See also Ch. II.   [3] *Ibid.*, pp. 108–110.

the same pace. "One lets credit expand a little faster than the others and loses gold; another lags behind and receives gold." [1] This does not remove the danger, however, for in time the internal drain depletes the reserves of banks, and credit must be curtailed.[2]

As stated above, Hawtrey's analysis refers specifically to cyclical movements of business, and the events he depicts will recur from time to time. His theory of gold movements is essentially classical, and primary importance is attached to changes in the prices of commodities resulting from fluctuations in the volume of credit.[3] But account is also taken of what he regards as lesser factors centering around the exchange market, and to these we shall turn presently.

[1] *Ibid.*, p. 110.

[2] It may be well to point out that Hawtrey assumes that changes in gold reserves will be effective in bringing about changes in the volume of credit outstanding in a very short time. Therefore, the central bank may concentrate its attention upon the position of its reserve and especially upon the situation in the exchange market. There is danger in such a policy, however, if the volume of credit outstanding is not immediately responsive to changes in reserves, i.e., the reserve is in part a hoard. To take an example: Suppose that a nation is expanding a little slower than other nations. It would tend to gain gold under Hawtrey's assumptions. Now credit would be expanding fast enough to bring the internal drain of which he speaks, and the reserve would eventually be drawn down. But no present action would be taken by the Bank at the time, except to encourage lending (which may not be effective as a policy), if its criteria of credit control are the reserve situation and the movements in the exchange market which are satisfactory as yet. Eventually, however, unless other circumstances intervened, the world-wide expansion would be halted by a crisis somewhere; and in that event rising discount rates elsewhere would draw gold from the country considered first. Thus the reserves would be drawn down by exports just at the time when the internal drain was becoming important. Hawtrey's later position that the index of prices is the guide which should be followed by the central bank is sounder, although even here the same difficulties are encountered. A production index, such as that used by the Federal Reserve Board, is an additional criterion of merit. But none of them meet the real situation, unless the rising interest rates affect commodity prices quickly.

[3] Chief reliance is placed upon the sensitiveness of traders carrying stocks of consumers' goods to discount rate changes. These traders have little working capital of their own, and are forced to adjust both domestic and foreign orders in accordance with the cost of bank advances. Particularly in the pre-war period, traders in other countries borrowed in London, and changes in Bank rate affected other nations quickly. See R. G. Hawtrey, "London and the Trade Cycle," *American Economic Review*, Supplement, vol. XIX, no. 1 (March, 1929), p. 71, and *The Art of Central Banking*, London, 1932, p. 381. It is recognized that new investment in capital goods is not affected readily by changes in interest rates, and the response of traders is very important as a consequence, as it breaks the circle and starts a new movement of the cycle. See *ibid.*, pp. 381–384. Hawtrey also states that changes in Bank rate will not affect prices of international trade goods as readily today as in

The present study is concerned largely with the concepts advanced by Hawtrey respecting internal drains, and the international flow of gold during the various phases of the business cycle. It seeks to give precision to the concept of lags, and to afford quantitative evidence concerning them. And it is desired to explore more fully the mechanisms of the foreign exchange market and the processes involved in adjustments of disequilibria in the balance of payments. In order to set the ground for this latter phase, it is desirable to return to the analysis presented by Hawtrey, as well as by Taussig, concerning the precise role played by gold movements in adjusting the balance of payments.

Hawtrey's analysis of this problem is undertaken in his chapter on the foreign exchanges, though it depends largely upon the concepts developed in earlier chapters. He supposes an expansion of credit in one country, which thereby brings an increase in the consumers' income and the consumers' outlay. In part this will increase the import of foreign goods, and the preexisting equilibrium of the balance of payments will be disturbed. The increased demand for imports means that the country's currency is less desired abroad, and there is a *virtual* depreciation of the unit. The exchanges become unfavorable, and gold is exported to pay for the excess of imports.[1] Equilibrium can be restored by contracting credit ". . . till the consumers' outlay is so reduced as no longer to attract an excessive share of the world's supply of foreign trade products."[2]

The processes are examined further in connection with a disturbance of equilibrium arising from the failure of a harvest, or the like. Such disturbances are reared on top of the normal course of cyclical disequilibria outlined above. The first effect of such a failure is a decline in consumers' income and in their outlay. In part this means a decrease in the purchase of foreign goods, and in part a decline in buying domestic commodities. Unless imports fall off immediately to the full extent of the decreased income

Ricardo's time because of the greater perfection of international markets and communications. But changes in demand for international goods may occur at a given price level, and equilibrium is thus secured. *Ibid.*, pp. 144 and 402–403.

[1] *Currency and Credit*, pp. 73–75.

[2] *Ibid.*, p. 74.

(which is unlikely) the foreign exchange market will be affected and the balance of payments will become adverse. But there will also be a diminution in the consumers' outlay on domestic goods, bringing repercussions to act upon the exchange market, and in time income and outlay on domestic goods will be equated, as will the income and outlay on foreign goods. When this has been accomplished, equilibrium is restored.[1]

The process is hindered by the reduction in the unspent margin, for these margins of consumers in the country experiencing the crop failure will be drawn down. Ultimately that part of the balances which is expended for foreign goods reaches the hands of foreign traders. These traders desire means of payments abroad for the increased imports (a relative increase in imports over exports is all that is essential for the analysis), and their increased demands cause the foreign exchange rates to become unfavorable. Bullion dealers take advantage of these rates to ship gold abroad, and sell drafts drawn against the foreign balances created, and in turn repay their advances secured from banks to purchase the gold. By this process the volume of outstanding credit and currency is reduced, and the unspent margin is diminished.[2]

The processes outlined are subject to modification from three sources. First, the fall in the prices of home-trade commodities will tend to divert demand to them from foreign goods. Second, if the disturbance is long-continued, there will be a diversion of enterprise from the production of domestic products, whose prices have fallen, to foreign trade products, whose prices do not tend to fall as rapidly. Third, capital movements will be affected. If profits fall off there will be more borrowing abroad (or a diminution of lending in the case of a creditor nation), and the severity of the impact upon goods and thence upon the exchange market is thereby lessened.[3]

Credit contraction will be furthered by the loss of gold, as bank reserves will be reduced, and this decline in the volume of credit will hasten the reduction of consumers' incomes and outlays, and a short cut to equilibrium is provided. To the extent that credit

[1] *Ibid.*, pp. 76–78.
[2] *Ibid.*, pp. 78–79.   [3] *Ibid.*, pp. 79–81.

contraction occurs, the pressure leading to gold export is diminished, and if the decline in the unspent margin is sufficient no specie need be sent abroad. It is probable that some gold would be lost, however, since less specie would be required for reserves because of the decline in the unspent margin.[1]

The analysis of the mechanism of trade adjustment which Hawtrey presents is inherently classical. It depends upon the prompt reaction of commodity prices to changes in the volume of deposits and currency. The size of the unspent margin must be sensitive to variations in rates of discount in the central money market, and money market rates must reflect changes in Central Bank rates quickly. The trader carrying stocks of goods is the focal point in the mechanisms set up, since changes in the volume of orders placed by him are assumed to be dependent upon the cost of bank credit. Alterations in the quantity of orders placed by traders affect the incomes and outlays of consumers and the amount of the unspent margin quickly. And once the equilibrium between consumers' incomes and outlays is disturbed, the forces set to work as a consequence will change the direction of the cycle without further impetus from Bank rates.

The mechanism set up may be criticized on several grounds. Central Bank control over the money market must be complete enough to cause market rates to respond to changes in Central Bank rates. Particularly during periods of long depression there may be little connection between these rates. And the close relation required may be broken at other periods by lack of securities in the Central Bank portfolio for open-market sales, or by movements of foreign balances which offset the efforts of the Bank. The central money market must also be relatively important in the general economy of the country, so that discount rate changes may affect business conditions in general, unless the Central Bank rate is effective directly upon borrowing throughout a nation. The position of the trader must be important enough to exert significant pressure upon the volume of bank deposits, if the cycle of expansion or contraction is to be broken. And the amount of borrowing by traders must be dependent upon fluctua-

---

[1] *Ibid.*, pp. 81–82.

tions of money rates. Where bank borrowing is dependent principally upon the willingness of bankers to lend, rather than upon small variations in profits of business men occasioned by changes in the cost of credit, the response of deposits to changes in Bank rates will be slow.

Unless the volume of bank credit is immediately sensitive to changes in rates, a movement of short-term foreign balances may offset the action of the Central Bank. Hawtrey recognizes that the flow of foreign balances may delay a contraction of credit, but he assumes that the rise in Bank rate will be effective, and that the same underlying equilibrium which would have been reached by allowing gold to flow out will be attained eventually. If the influx of balances is precisely equal to the gold exports which would have taken place, there is no delay in the attainment of equilibrium. If, however, the rise in Bank rate attracts foreign balances to a degree which more than offsets the adverse balance of payments, the Central Bank would hesitate to keep such a high rate, and the credit contraction would be delayed. Moreover, the influx of gold would tend to make the rate ineffective, unless open market sales of securities are made.[1] But the credit contraction assumed by Hawtrey may not even occur. The influx of balances may be so large as to nullify completely the action of the Central Bank, and the net effect may be an increase in the unspent margin for short periods.[2] While ultimately contraction of credit may occur, the expansion phase of the cycle may continue for periods of months or possibly years before the increased cost of bank credit or the unwillingness of bankers to make further loans brings the phase to a close. Where a nation is borrowing heavily abroad short-term money market rates may exert relatively little influence, and the expansion phase of the cycle will be extended despite high rates. In periods of depression lowering Bank rate to stimulate traders

[1] *The Art of Central Banking*, pp. 415–418. It may be pointed out that the credit extended through the foreign balances may also defeat the aim of the Central Bank. It is not essential that gold accompany the flow of funds in order to affect discount rates in the money market in a manner not desired by the Bank.

[2] Hawtrey recognizes that investment in new enterprises responds somewhat slowly to changes in Bank rate. If money market rates are held down closely to the old level by entrance of foreign balances, the effect of the Central Bank policy upon new investment may be almost negligible for a considerable period.

may bring such an exodus of foreign balances as to decrease the unspent margin, and thus further prolong the downward phase of the cycle.[1]

(5) Writers on foreign exchange have contributed an important body of literature on the question of gold movements. Starting with an examination of the mechanisms of the exchange market, forces at work within the market itself and the foreign exchange banks have been depended upon for the adjustment of disequilibria in the balance of payments, and the price-specie flow analysis has been neglected, or, in some cases, explicitly denied. The somewhat scattered ideas of previous writers were brought together by Goschen, and a brief statement of his views is presented below. The influence of Goschen and others upon economists has been important, and gold movements are regarded by Laughlin as results of capital movements solely. On the other hand, Taussig,[2] Whitaker, and Young have recognized the importance of adjustments by exchange market forces where disequilibrium is not quantitatively important, but insist that where the departure from equilibrium in the balance of payments is great or protracted that the price-specie flow mechanism comes into play.

The elements which determine fluctuations in exchange rates according to Goschen are summarized by Angell as follows: "Currency disturbances apart, they are: (1) the balance of indebtedness, which, however, will cause fluctuations only within the specie points, (2) the relative rate of interest in the two countries, (3) the state of credit in each, and (4) political disturbances; or some combination of the four." [3]

---

[1] The importance of the movement of foreign balances in the post-war period has been pointed out many times. See, e.g., the articles by Jacob Viner and John H. Williams in *Gold and Monetary Stabilization* (Lectures on the Harris Foundation, 1932), edited by Quincy Wright, Chicago, 1932; John Maynard Keynes, *A Treatise on Money*, London and New York, 1930, Chs. 21 and 36; and Feliks Mlynarski, *The Functioning of the Gold Standard* (Memorandum Submitted to the Gold Delegation of the Financial Committee of the League of Nations), Geneva, 1931, Ch. II.

[2] See above, pp. 9–10.

[3] Angell, *op. cit.*, p. 139. The first writer, according to Angell, to point out the significance of discount rates, acting on the market for bills, in determining the flow of specie, was H. D. Macleod, *The Theory and Practice of Banking*, London, 1855–1856, vol. II, p. 344. Macleod also shows that the two ways to correct an adverse

Goschen goes on to explain that usually the balance of indebtedness on account of goods and the rate of interest are forces which work *simultaneously, but in opposite directions*:

> The adverse balance of trade will, as far as its power extends, render the bills on the country which is most in debt difficult of sale, and tend to compel it to export specie; whereas the high rate of interest, which is contemporaneous with a drain, *or the prospect of a drain*, of specie, will revive a demand for bills on this same country, and enhance their value in other quarters; for there will be a general desire to procure the means of remitting capital to that market where it commands the highest value.[1]

The analysis is completed by showing that the natural course of events would bring about a rise in the rate of interest. "The abstraction caused by the bullion shipments will of itself tend to raise the rate; and banking establishments will in their own interest (which will be identical with the interest of the public) accelerate this result as far as it lies in their power." [2] The action of central banks is, of course, directed powerfully toward the same end. Goschen does not, however, in his analysis explicitly suggest that the rise in interest rates precedes the drain of specie, or that such a rise may avert a threatened loss, although this possibility is implicit in his reasoning.[3]

It is worth while to note another point developed by Goschen with regard to the significance of demand as opposed to long bills in determining the movement of gold. He maintains that

exchange, where the currencies are not depreciated, are by exports of produce and a rise in the rate of discount. *Ibid.*, vol. I, p. 431 (fifth edition, London, 1893).

The analysis is indicated in a fragmentary way by S. J. Loyd somewhat earlier. See below, pp. 31–32.

[1] G. J. Goschen, *The Theory of the Foreign Exchanges*, London, 1861, p. 127. (The italics are mine.) I have used here the reprint of the third edition published by Wilson (London, 1894). Goschen notes that ". . . when in 1861 the excessive indebtedness of England to America (the result of her importations of cotton and corn) lowered the price of English bills in New York, and rendered specie remittances to the States inevitable, the high rate of interest in England which this situation had brought about, was so attractive to continental bankers, that they drove up the price of bills upon this country to specie point, and were finally induced even to resort to bullion remittances." *Ibid.*, pp. 127–128. Cf. also p. 131.

[2] *Ibid.*, p. 132.

[3] The failure to suggest this possibility may be explained by Goschen's conclusion that the Bank of England's rate of discount is merely an indication of market rates, and that the Bank does not have the power to control the market rates. See *ibid.*, pp. 133–135.

it is the price of short bills, not of those which have some time to run, which determines the course of bullion shipments. Most of the primary elements of value affect long and short bills equally; but the rate of interest and the question of credit exercise an additional influence upon the former, and so modify the fluctuations in their price as to render them unreliable as indications of the currents of gold.[1]

While this is undeniably true for domestic bills, the point cannot be pressed too far when dealing with the foreign exchange market under present banking conditions, for there is a strong underlying tendency for all bill and exchange rates to move together. Let us assume a condition in which England has an adverse balance, and bills on London in New York are consequently plentiful. Exchange rates would be near the gold import point in the latter center. The prospect of bullion exports would tend to bring advances in the rate of discount in London and thereby widen the spread between the prices of short and long bills. Bankers in New York would be induced to hold their long bills instead of discounting them in London as usual. They would then be unable to draw sight drafts as cheaply, since their London balances would tend to decline.[2] The effect would be a rise in the price of demand bills on London in New York, and a rise in the price of long bills in London.

(6) An analysis similar in many respects to that of Goschen in its emphasis upon the exchange market is presented by Laughlin. He denies the price-specie flow mechanism of the classical school, and maintains that gold movements are the result of disturbances in the normal equilibrium of trade (or payments) and that they

---

[1] *Ibid.*, pp. 88–89. See also pp. 135–140.

[2] See Franklin Escher, *Foreign Exchange Explained*, New York, 1917, pp. 50–53. The important part played by bankers is stressed by Furniss. "In each of the transactions we have discussed there has been involved a decision on the part of foreign bankers either to cease to borrow on the London market or to lend in that market. The purchase of sterling long bills for investment instead of for discount, the discontinuance of the practice of drawing sterling finance bills, the closing out of loans already contracted, all indicate a decision by foreign bankers to withdraw from the London discount market as borrowers; the remittance of demand sterling to London bankers for the purpose of making loans in London brings foreign bankers into the discount market of that city as lenders." Edgar S. Furniss, *Foreign Exchange*, Boston, 1922, p. 352.

can never be the cause of changes in the trade balance.[1] This position is inherent in his general refutation of the validity of the
quantity theory of money.

His doctrines rest in part upon his propositions with regard to
the function of gold in the price structure of a country. Gold can
affect prices only as it becomes a part of the general media of exchange,[2] but he extends this notion to cover modern conditions
where gold is used largely as bank reserves. An excess of reserves
does not, however, necessarily bring about the issuance of more
credit, for

in legitimate banking, loans are made because of satisfactory collateral or
actual transfer of goods, and not merely because reserves are high. To be
sure, if reserves rise, rates of interest will fall and new loans are possible; but
merely because a bank can loan, it does not follow that it does loan. . . . If
gold is pouring in beyond the needs of banking safety, the banks get rid of it,
just as of any other asset which does not pay a return.[3]

Having repudiated the classical approach to prices and international trade adjustments, he proceeds to set forth his own views.
He points out that it is the movement of interest rates rather than
the movement of prices of commodities which determines the
flow of gold. Under modern conditions the exchange of securities,
and of other services, must also be considered. The amount of
gold moving under these circumstances becomes very small in
comparison, and can have very little or no effect upon prices of
goods. His position with regard to the significance of interest
rates is so important that it merits the following full statement:

The reason why balances in favor of a country may not be paid in gold is
due to the possibility of investing those balances at a higher rate of interest in
a foreign country than can be obtained at home. The relative rates of interest
have an influence even wider than that upon the movement of balances. It is
the rate of interest upon sound international securities, as well as the rate in
the loan market, which determines whether credits due, for instance, to the
United States shall be left abroad or brought home in the form of goods or
specie. The general accounting in foreign trade must include the operations
of loans, and the movement of capital for investment from one country to another. Indeed, the rate of interest is behind the movement of securities men-

---

[1] J. Laurence Laughlin, *The Principles of Money*, New York, 1903, Ch. X, esp.
pp. 369–377.
    [2] *Ibid.*, p. 371.   [3] *Ibid.*, p. 387.

tioned above. The purchase of securities is, of course, one form of investing capital. Therefore, in determining the causes affecting prices and the movement of gold in international trade, it must be kept in mind that the relative rates of interest in the trading countries will influence the passage of loanable capital to and fro, thereby acting as a factor in adjusting merchandise credits and debits, and seriously affecting the transmission of gold. Instead of gold being the originating cause of new exports and imports, as once generally held, it is the very last thing to move; and even then merchandise balances may be entirely reversed by changes in the rates of interest in New York or foreign centres which may cause capital to flow from the creditor to the debtor country. The recognition of the force exerted by the rate of interest on the movement of loanable capital gives the final *coup de grace* to the old theory, which based its change of general prices upon the international movement of specie. The order of events is quite the other way: relative prices cause exports and imports of goods; and the shipment of gold is not necessarily made even to cover balances of merchandise. If gold moves, it goes not merely because of the account in goods and securities, but of the investment of international capital.[1]

Laughlin also makes his position clear with respect to gold movements at a time of commercial crisis:

It might be said, however, that there could be a temporary change in the value of gold in one country, due to violent trade convulsions, which did not extend to other countries; and that the value of gold would be different for a time in that country from its world value. If so, prices would be for a period elevated or depressed, and gold would be moved to restore the equilibrium. Could there be such a temporary scarcity of gold in one country? A case in point may be cited, in the history of the panic of 1893 in the United States, when there existed what was called a "money famine," and when gold was imported.

Such a situation would, even before the event, show itself in the rate of interest, and gold could be imported within a week. Granting the temporary scarcity of gold in such an emergency, the time required for the shipment of gold could not be long enough to change the general level of prices; in truth, it would be usually found that prices had fallen before the money famine disclosed itself. This exceptionally high value set on gold, in 1893, was due to temporary business and banking conditions, and to the liquidation of obligations. In this case the sudden demand for gold was disassociated from the movement of prices; the increased estimate put on gold in the panic came after, and so could not have been the cause of, a previous fall of prices.

To be sure, a fictitious rise of prices due to abnormal credit might temporarily give to gold a lower value within a country than it possessed elsewhere in the world; but these conditions bring their own overturn in such vengeful fashion as to show that it is at variance with the natural principles of price-making.[2]

---

[1] *Ibid.*, pp. 382–383. [2] *Ibid.*, p. 390.

It is worth while to note at this time an article of Professor Whitaker's [1] which subjects this analysis of Laughlin to careful and penetrating criticism. With much of this article we need not concern ourselves, but the author's view that the rate of discount acting on the movement of securities is "temporary and of strictly limited scope" [2] is worth attention. If the excess of imports into a country is large, or if it continues for a long period, the discount rate becomes an important influence on the flow of gold. The influence of such rates is confined largely in his view to the correction of normal season flows. Important differences between imports and exports can be settled only by the movements of gold.

Whitaker also shows that for the years 1890–1903, with the exception of 1897, net changes in the annual movement of gold produced similar changes in the excess of merchandise exports over imports for the succeeding year in the case of the United States. While he does not link this notion definitely to cyclical fluctuations, as Hawtrey does, such reasoning is possible upon the analysis which he does present.[3]

A classification of the various types of gold movements which is admirable in many respects is presented by Whitaker. He divides them into (1) producers, (2) commercial, and (3) financial or money market shipments.[4] The second type are due to trade

---

[1] A. C. Whitaker, "The Ricardian Theory of Gold Movements and Professor Laughlin's Views of Money," *Quarterly Journal of Economics*, vol. XVIII, no. 2 (February, 1904), pp. 220–254.

[2] *Ibid.*, p. 232.

[3] *Ibid.*, pp. 249–253. The following passage is significant: "Dynamical influences (of production and commerce, entirely independent of the Ricardian specie forces) led our excess of exports to increase generally through the period. But, while pursuing this general line, the excess swings from side to side along the course. Let us adopt the term 'invisible balance of trade' for the sum of all the accounts other than the real 'balance of trade' (in goods), which enter the 'total balance of indebtedness.' Now, whether the 'invisible balance of trade' remains constant or shows an increasing excess of credits or of debits, the 'real balance of trade' in adjusting itself to this movement is certain to oscillate about its necessary course. Its necessary course is that which will compensate for an excess of credits or of debits to an equality (or, more precisely, to that position which will occasion that gold movement which is requisite to preserve the national quota). Along the dynamical path of our foreign commerce at one time our exports swing too far in excess even of our expanding debits, then again they swing below them. Thus they produce flows of gold, at one time to the country and again from it." *Ibid.*, pp. 252–253.

[4] *Ibid.*, p. 221.

or service balance items, while the last are due to differences in interest rates, and may arise from direct loans of gold (on collateral) or from the purchase of floating securities. The difference between commercial and financial shipments is set forth clearly: "A commercial shipment of gold always serves to discharge a previous international debt, but an initial financial shipment creates a debt and looks to a future and early return shipment of gold." [1]

Whitaker maintains that adjustments through borrowing operations on floating securities are necessarily only temporary. But they serve to diminish gold flows. Professor Allyn Young also made a somewhat similar analysis.[2] Taking the case of a country experiencing a cyclical rise of prices, he points out that under classical analysis an export of gold would be expected. But interest rates also advance along with prices, which tends to diminish lending by the country, or increase loans to it, or both. This will tend to diminish or delay specie exports. "The delay may even continue until the balance of trade has turned in the other direction, so that gold shipments become unnecessary." [3] Thus movements of capital soften the shocks attendant upon the movement of gold. And he also points out that bankers do not move gold in order to take advantage of high rates of interest, but that loans in this case merely mean that the gold reserves of other centers are put at the disposal of the borrowers.[4] He does not show why the trade balance should be affected ultimately, although he undoubtedly had in mind the influence of high interest rates upon prices and production.

Analysis of the foreign exchange market has directed the attention of later writers to the importance of capital movements, and brought greater recognition to the role played by interest rates in the adjustment of disequilibria in the balance of payments. Refinements in the discussion of cyclical influences have resulted from the work of the foreign exchange writers. The importance

---

[1] *Ibid.*, p. 224.

[2] His views are to be found in R. T. Ely, T. S. Adams, M. O. Lorenz, and A. A. Young, *Outlines of Economics*, New York, 1908. I have used here the fourth edition (New York, 1923).

[3] *Ibid.*, p. 349.   [4] *Ibid.*, p. 350.

of their contribution is shown further in the following section, where certain contemporary writers are considered.

(7) The influence of the writers on foreign exchange is shown clearly in the work of Angell. Transfers of purchasing power through the bill market, and the resultant effects upon the volume of bank deposits and loans and upon the level of commodity prices, replace gold movements as the principal agent for the adjustment of disequilibria in their view. In many cases the adjustment occurs within the limits of the gold points, and even where specie is moved, the flow is regarded as a consequence of conditions in the banking systems already set up by the transfer of purchasing power.

Angell [1] considers both short- and long-period adjustments, and he relates the problems to the fluctuations of the business cycle. The mechanisms of adjustment are illustrated by the case of the flotation of a large international loan. The lending country (designated as country A) is now under the necessity of transmitting the proceeds of the loan to the borrowing nation (country B). It must therefore increase the relative volume of remittances it has been making to foreign countries.[2]

He then proceeds to examine the forces relating to the holding of bills of exchange by bankers. The stock of bills is built up by discounting drafts for exporters and the volume of deposit credit is thereby increased, while the sale of bills to importers reduces the stock of bills and decreases the purchasing power of the country.[3] The net effect of these two sets of forces is that while the holdings of the banks vary, they may be said ". . . to follow paths which we may describe as 'normal'." [4] This "normal" would allow for seasonal and cyclical movements of business, and secular tendencies which are primarily domestic in origin.

In a later article [5] Angell examines the mechanisms of the bill market in more detail. The manner of financing foreign trade

---

[1] *Op. cit.*, Ch. XVI.   [2] *Ibid.*, pp. 402–403.
[3] *Ibid.*, p. 403.   [4] *Ibid.*, p. 404.
[5] "Equilibrium in International Trade, The United States, 1919–1926," *Quarterly Journal of Economics*, vol. XLII, no. 3 (May, 1928), pp. 388–433.

transactions, the policy of commercial banks, and the actions of the foreign traders affect the volume of purchasing power. In some cases imports *may* bring an expansion of purchasing power in the importing country. If the imports are paid for by the purchase of a foreign currency draft from a bank, the bankers' foreign balance is drawn down.[1] But the bank gains an equivalent amount of domestic funds, and its lending position is improved thereby.[2] Moreover, the importers or the buyers of the goods may borrow in order to purchase their drafts, and while ultimately repayment of the loan extinguishes the volume of purchasing power created by it, the aggregate loans outstanding may increase if there is overlapping of individual loans.[3] Where payment for imports is effected through the drawing of a draft upon the importer or his bank, the foreign balances of the exporter or his bank will be increased after acceptance and sale of the bill.[4] Both the assets and the liabilities of banks in the importing nation are increased.[5] An expansion of purchasing power may occur if banks extend additional loans during the interim between acceptance and maturity of the draft.[6] The effect of exports upon the volume of purchasing power is also examined in detail. It is important to note that *either* imports or exports may increase the volume of deposits.[7]

Returning to the case of the flotation of a foreign loan, Angell states that the agents of the borrowing country will enter the exchange market to purchase bills in return for the local currency or deposits paid to them by the investors.[8] "This new demand for bills will accelerate the rate at which the banks' holdings are depleted." [9] The average size of holdings will *tend* to fall as a result. This might continue until the store of bills was wiped out, but certain correctives are called into play before that will result.

First, the price of bills will rise, and exchanges will become un-

---

[1] *Ibid.*, p. 399. [2] *Ibid.*, pp. 400–401.
[3] *Ibid.*, p. 406. [4] *Ibid.*, pp. 399–400.
[5] *Ibid.*, p. 401. [6] *Ibid.*, p. 403.
[7] *Ibid.*, pp. 419–420.
[8] *The Theory of International Prices*, pp. 404–405. Angell does not consider here the possibility that the borrowers will draw drafts against the lending banks, and sell them in the exchange market of the borrowing country.
[9] *Ibid.*, p. 405.

favorable. This will stimulate the export of goods and restrict imports, within the limits of the specie points, in accordance with Mill's analysis. Further, it will encourage the drawing of finance bills, and stimulate other speculative operations.[1] Secondly, there will be a flow of gold abroad, both on account of the exchange position, and because of commitments of bankers which cannot be easily met by the purchase of bills. If protracted, these exports will have far-reaching effects, but the amount is usually so small that it has little effect on the adjustment in the balance of payments.[2] Thirdly, any large disturbance will bring increases in the discount rate in the lending country. This will operate to check the demand for bills, and will lead to a return of gold.[3]

The rise of rates is due to the loss of gold, and also to the loss of purchasing power produced by the decline in bill holdings.

> The decline . . . will create a demand for more bank credit to replace that part of the country's purchasing power which has (in effect) been exported. But to yield to these demands would be tantamount to inflating credit. The physical volume of business in the country has not increased, and there is hence no legitimate excuse for an expansion of loans.[4]

The above correctives are of limited duration, however. More important are the effects produced by changes in purchasing power. It will be recalled that the original purchaser of foreign securities has spent his income, and cannot, therefore, purchase domestic commodities (or more likely domestic securities). As a result there will be a decline in the purchasing power expended in the lending country, either directly or indirectly. This tend-

---

[1] It will be observed that in essence these are loan operations, and that the loans are to country A (the original lending nation). In other words, the lending area is extended to other countries, for the moment at least.

[2] *Ibid.*, pp. 405–406. Unless there are hoarded reserves, the repercussions upon credit are tremendous where banking reserves are attenuated as in nations of the western world. This is Hawtrey's thesis, of course.

[3] *Ibid.*, p. 406.

[4] *Ibid.*, pp. 406–407. In the case of disequilibrium due to an excess of imports, Angell sets up three corrective forces: (1) a premium on bills, and possibly gold exports, (2) a curtailment of credit by banks whose reserve ratios are declining, with gold exports also possible, (3) a decline in demand for domestic goods as consumers divert expenditures to imports, an increase in the total amount of goods and a smaller rise in purchasing power, both leading to a fall of prices. See "Equilibrium in International Trade," *op. cit.*, pp. 407–408.

ency will be accelerated by the loss of gold and securities from banks, and the consequent rise of discount rates. The *absolute* effects will vary in accordance with the position of the country in the business cycle; in some cases a mere retardation of the upward movement of prices will be the result. The final adjustments will take place through changes in the commodity balance of trade, and equilibrium will be restored. More goods will be exported and fewer imported, though some minor changes may result in other items of the international accounts.[1]

The conclusions may well be summed up in Angell's words:

> In the first instance, the disturbance leads to a movement of the foreign exchanges. If it is small and temporary, this adjustment alone may suffice, for it will accelerate or retard the rate at which bills of exchange are currently offered in the market. If that is inadequate, gold will flow; and discount rates, especially in the gold-exporting country, will be affected. If even this fails to correct the situation, finally an alteration will take place in the underlying conditions that govern the general course of international exchange itself. The volume of purchasing power in circulation in the creditor country will be built up, in consequence of the increase in the banks' holdings of bills. In the other country it will be reduced, through the decline there in bill holdings. These changes in turn will operate upon the corresponding general price levels. The latter effect is often, though not always, strengthened by alterations in the discount rates. The movements in prices will then influence the commodity balance of trade, in ways long familiar, and will continue to do so until the change in the commodity trade has become great enough to offset and correct the original disturbance in the balance of international payments.[2]

He continues by pointing out that temporary disturbances are adjusted by the immediate and short-time influences, but that large and enduring disturbances can only be corrected by an alteration in the trade balance itself. He shows that the classical theory took little account of these temporary correctives, except for the limited effect of exchange rate fluctuations on commodity movements, and that their main reliance was on movement of gold, which exercised a quick and direct influence on prices. He then restates his own interpretation of the role of gold flows:

---

[1] *The Theory of International Prices,* pp. 407–409. See also pp. 409–411 and Appendix B for an analysis of empirical data.

[2] *Ibid.,* p. 413.

The doctrine which has been set up here, on the other hand, regards gold flows as an essentially short-run phenomenon, usually of temporary importance alone. It bases its more long-run conclusions on the effects which significant changes in the relation between the demand and supply of *bills of exchange* produce upon the total volume of purchasing power in circulation. It rejoins the classical line of analysis only at the point where general price changes begin to appear.[1]

Angell is even more explicit when he considers the case of Canadian borrowing. He holds that the change in Canadian prices was brought about by an increase in bank deposits, and that gold imports or changes in the "outside reserves" were not the initial causal elements. Alterations in the "outside reserves" were due to lags in the response of international trade to fluctuations in the volume of loans, and they did not have a direct or immediate effect upon conditions in Canada.[2] "The correction, in the Canadian case, did not come from the effects of gold flows, or of changes in outside bank balances. It came from the effects of the original and prior increase in Canadian bank deposits."[3]

An additional element in the process of adjustment of disequilibrium in the balance of payments is suggested by White.[4] The volume of international trade will be affected by shifts in demand schedules for groups of commodities. The transfer of purchasing power may be made through gold movements[5] or short-term capital flows, and the movement of sectional price levels may either aid or impede the subsequent adjustment. In the case of a foreign loan, it would be expected by White that the borrower would use a large portion of it in purchasing imported commodities, and there would be an increase in the demand schedules for such goods,

---

[1] *Ibid.*, p. 414. See also Ohlin, *op. cit.*, esp. Ch. XVIII.

[2] Angell, *The Theory of International Prices*, pp. 170–174, and Appendix B, pp. 505–510, and his review of Viner, *Canada's Balance of International Indebtedness, 1900–1913*, in *Political Science Quarterly*, vol. 40, no. 2 (June, 1925), pp. 321–322.

[3] Angell, review of Viner, *op. cit.*, p. 322.

[4] *Op. cit.*, pp. 17–35. See also Roland Wilson, *Capital Imports and the Terms of Trade*, Melbourne, 1931, Ch. IV, and Ohlin, *op. cit.*, Chs. XX and XXI.

[5] *Total* gold movements are important, as purchasing power is transferred by specie imports even though exports may reduce the *net* import to a negligible quantity within a short period. Banks would be led to increase loans and deposits as a result of the import, and changes in demand schedules might result and adjustment in the international balance of payments achieved which would permit the export of the specie within a short period. See White, *op. cit.*, pp. 30–32.

quite apart from any increase in the quantity taken because of the effect of the transfer of purchasing power upon sectional price levels. The loan is thus taken in the form of goods without any changes in the export-import price relation, and the classical price specie-flow mechanism is unnecessary for adjustment to occur.

(8) The chief proponents of the Currency Principle,[1] Lord Overstone, Robert Torrens, and George W. Norman, applied the unilateral Ricardian theory of gold movements to the monetary problems of the day. They held a rigid form of the quantity theory of money, and argued that a bank note currency should fluctuate in the same manner that a pure metallic currency would if equilibrium of prices was to be achieved.[2] Outflows of specie would bring adjustment of disequilibrium in the balance of international payments provided the note issue was contracted concurrently with gold exports.[3] The Bank of England was

[1] The discussions of the Banking-Currency School controversy by T. E. Gregory, *Select Statutes & Reports Relating to British Banking, 1832–1928*, London, 1929, vol. I, pp. xvl–xxiv, and by A. E. Feavearyear, *The Pound Sterling*, Oxford, 1931, pp. 243–254, have been very useful.

[2] S. J. Loyd (Lord Overstone), *Reflections Suggested by a Perusal of Mr. J. Horseley Palmer's Pamphlet on the Causes and Consequences of the Pressure on the Money Market*, London, 1837, pp. 15–16, and Robert Torrens, *A Letter to the Right Honourable Lord Viscount Melbourne on the Causes of the Recent Derangement in the Money Market and on Bank Reform*, London, 1837, p. 29.

[3] Loyd argued that such contraction was essential: "A decrease in the amount of bullion requires a diminution of the circulation. If the drain [of bullion abroad] has arisen from peculiar or local circumstances, and whilst the general state of the exchanges is not unfavourable, a very slight contraction of the circulation will cause an influx of bullion and thus remove the evil; but if the circumstances of the case should be misunderstood, and the drain, though apparently arising from internal or local causes, should in reality be connected with an improper state of prices and a consequent general tendency to send gold abroad (no improbable supposition); in that case any delay in effecting the necessary contraction of the circulation will only tend to defer and prolong the crisis without diminishing its intensity, and thus the mass of suffering may be augmented, whilst by the temporary abandonment of sound and consistent principles the very basis of the circulation may be exposed to danger." *Further Reflections on the State of the Currency and the Action of the Bank of England*, London, 1837, pp. 32–33.

Later Loyd pointed out that a rise in the rate of interest by the Bank would arrest an outflow of bullion. This is not only powerful, but quick acting. He refers to such a case in April, 1847, saying that as a result of the rise in the value of money there ". . . was an *immediate* correction of the exchanges — the drain of bullion at once ceased — the gold actually shipped for America was relanded — policies upon further shipments were cancelled — and the bullion previously sent to the Continent of

attacked because it did not vary its issue in accordance with this principle.[1]

Defenders of the Bank pointed out that it had no control over the issues of country banks, and held that these local bank notes were largely to blame for the inflation of the currency.[2] In emphasing the importance of deposit credit, the Banking School made an important contribution, although some of the adherents of the School were clearly erroneous in maintaining that there was an automatic "reflux" of notes issued in connection with bank loans and that the temporary expansion of the currency would not affect prices.[3]

Fullarton's analysis of specie flows is suggestive, even though it is confused and at times contradictory. He denied that gold movements could be brought about by excessive note issues or by a relatively high price level. Exports of specie were caused by an unfavorable balance of payments.[4] But he admits later that "speculation and over-trading" may bring gold flows.[5] He suggests as other causes of drains demands of foreign governments for specie for coining, demands for hoarding abroad, over-investment in foreign securities, and increased imports of goods because of a harvest failure or some similar disturbance.[6]

Europe began forthwith to return to this country." *The Petition of the Merchants, Bankers, and Traders of London, against the Bank Charter Act: With Comments on Each Clause*, London, 1847, p. 7. Here Loyd is concerned with movements at the *time of crisis*. He is not considering the effect of interest as a cost to business, which ultimately reacts on the "trade" balance.

[1] See, e.g., S. J. Loyd, *Reflections Suggested by a Perusal of Mr. J. Horseley Palmer's Pamphlet on the Causes and Consequences of the Pressure on the Money Market*, p. 36; T. Joplin, *An Examination of the Report of the Joint Stock Bank Committee*, London, 1836, pp. 66–68; David Salomons, *A Defence of the Joint-Stock Banks; An Examination of the Causes of the Present Monetary Difficulties and Hints for the Future Management of the Circulation*, London, 1837, p. 6; Torrens, *op. cit.*, pp. 23–27; and W. Bennison, *The Causes of the Present Money Crisis Explained, in Answer to the Pamphlet of Mr. J. Horseley Palmer; and a Remedy Pointed Out*, London, 1837.

[2] See, e.g., J. H. Palmer, *The Causes and Consequences of the Pressure upon the Money-Market*, London, 1837, p. 38.

[3] John Fullarton, *On the Regulation of Currencies*, London, 1844, p. 79. I have used here the second edition, London, 1845. And see Angell, *The Theory of International Prices*, pp. 75–77.

[4] Fullarton, *op. cit.*, pp. 130–131.

[5] *Ibid.*, p. 161.

[6] *Ibid.*, pp. 153–161.

The level of prices is determined by the volume of bank credit and not by the quantity of currency, and the Bank cannot control specie flows by regulating the amount of the currency. Its power of control through regulation of deposits or notes is limited. In some cases the specie which is exported comes from "hoards in the hands of the banks or the public, and the movement will not affect prices at all." [1] In the case of drains due to foreign government demands or to panic conditions it is impossible to stop their progress, and it is undesirable to hinder flows which pay for imports of grain necessitated by a failure of the harvest at home. The Bank of England should, therefore, hold a sufficient stock of bullion to meet such demands, until other fortuitous circumstances arise to bring the gold back again.

Drains due to over-trading are of a different nature, and the Bank should have no compunction in dealing with them. But, in the main, it is impotent, for these come at the end of a period of speculation, when the collapse is imminent. [2] He shows that an influx of bullion is to be found in most cases when the circulation of notes in the country is large. He continues:

> The case is nothing more than this, that the same circumstances which conduce to a rise of the exchanges and an influx of gold, denote also generally the existence of an active state of internal industry, a high scale of production and consumption, and every condition requisite to an enhanced employment and demand for money. On the other hand, when a fall of the foreign exchange and a drain of gold come in the train of a period of commercial excitement and speculation, their advent is commonly the signal of a collapse already commenced, an indication of overstocked markets, of a cessation of the foreign demand for our productions, of delayed returns, and, as the necessary sequel of all these, of commercial discredit, manufactories shut up, artisans starving, and a general stagnation of industry and enterprise. [3]

Fullarton employed this concept merely to show that fewer notes were needed in circulation when business was slack, and to prove that the state of the note issue did not determine gold flows. He also maintained that the Bank could be effective in con-

---

[1] *Ibid.*, pp. 71 and 140–143. Fullarton even maintains that variations in the output of gold mines do not affect prices, since excessive supplies are absorbed by the hoards, and diminutions in output serve to call forth gold from the hoards.

[2] *Ibid.*, pp. 150–153.

[3] *Ibid.*, pp. 126–129.

trolling the state of credit and currency only occasionally. For normally the country was not dependent upon the Bank.[1]

By this time the nature of the internal demand for money at the time of a panic was clearly understood. Demands for coin and notes at such times were due to fear, and money was wanted for hoarding — not spending. Therefore the increased amount in "circulation" could not affect prices as it would in normal times.[2] And the proper procedure for the Bank of England to follow was distinctly shown. In order to stay the panic, loans were to be freely granted, according to Torrens, for the internal situation was the more precarious. But this did not excuse the Bank for failing to contract earlier when gold was being exported.[3] The terms of Bagehot's thesis are evident here, although the increase in the rate of interest charged by the Bank, which is an integral part of the latter work, is not stressed until Overstone's treatises after the crisis of 1847.

Out of the controversy over the Bank Act of 1844 there emerged important contributions to theory. The Currency School theorists applied the Ricardian doctrine of specie flows to the monetary situation, and pointed out that the correct procedure for banks was to curtail note issues early in a period of expansion. Otherwise the trade balance would become unfavorable, and gold would flow out, bringing panic conditions. If action were taken early, the disequilibrium could not become large, and the loss of gold would be thereby reduced in volume. The restriction of currency would affect prices and the trade balance, and restore equilibrium in the international accounts. By the Act of 1844 the Ricardian principle was established, and the Bank was forced to act in the manner prescribed, since the amount of expansion of note issue was limited by the bullion stock at the Bank. It took many years

---

[1] *Ibid.*, esp. pp. 137, 150, and 166.

[2] An anonymous writer in 1848 draws an analogy between the demand for money during a panic and the actions of a crowd as it struggles from a burning theatre. The progress is actually slower than usual; and similarly money is actually scarcer than usual during a panic. *The National Distress: Its Financial Origin and Remedy*, London, 1848, p. 35. See also Joplin, *op. cit.*, pp. 81–82.

[3] Torrens, *op. cit.*, p. 43. Joplin maintains that the Bank reversed its policy in 1825, and began to grant credit freely after he published a letter in the *Courier* showing the correct procedure. *Op. cit.*, esp. p. 86.

of agitation and criticism on the part of Bagehot [1] and others to force the Bank to recognize its responsibilities, and to take action *early* in a period of prosperity, and vary its rate of discount in accordance with the state of the Reserve.

On the other hand, the critics of the principle had pointed out that there did not exist the intimate connection between gold and prices which was presumed by the Currency School. In part this was due to the development of deposit credit, in part to the lack of control by the Bank over the business of the country, and in part to the fact that gold flows did not come from circulation but from "hoards." Moreover, it was asserted that the business of the country should not be hampered because of gold flows arising from causes other than over-trading and speculation. Therefore, bullion reserves should be elastic to some extent to allow for irregular and spasmodic gold flows traceable to causes other than relative states of price levels. While the Act itself did not admit of great latitude in that respect, the effects of such flows, as they became easily recognizable, were discounted by the market.[2]

(9) The influence of the Currency School is shown clearly in the treatment of specie flows by de Laveleye and Juglar. Gold movements, according to de Laveleye, are dependent largely, in the first instance, upon the exchange market and the rate of discount.[3] He followed Overstone and Goschen in showing that a high rate of interest will bring specie back to a country which has lost it. He also showed that internal drains might be important, especially at a time of crisis.[4] But at times the internal flow might

---

[1] Walter Bagehot, *Lombard Street: A Description of the Money Market*, London, 1870, esp. p. 48. I have used here the sixth edition (London, 1875). Bagehot notes that periods of internal panic and external drain commonly occur at the same time. The first must be remedied by ready loans, and the second by raising the discount rate to attract foreign gold. *Ibid.*, p. 56.

[2] *Ibid.*, pp. 319–320.

[3] Émile de Laveleye, *Le marché monétaire et ses crises*, Paris, 1866, esp. p. 161.

[4] He says, for example: "The great crises have almost always attained their maximum intensity in the autumn, because then the exportation of money toward the country comes to join that toward foreign countries." *Ibid.*, p. 119. And also: "In England toward the end of the year, attracted by a high interest and by the low price of securities, money arrived from the provinces and from all parts of the continent with a rapidity which exceeded all expectations. . . ." *Ibid.*, p. 82.

supply the gold sent abroad.[1] The latter type of movement is not treated specifically in connection with cyclical fluctuations of business, and it is probable that seasonal flows only are thought of, but the analysis suggests that of Professor Young in connection with the United States.

Juglar's analysis is more complete. Like de Laveleye, he stresses the actions of the foreign exchange market. He attributes the precipitation of a crisis to an export of gold. But the concept is extended to cover other phases of the trade cycle. In the earlier stages prices tend to rise, the balance of trade tends to become unfavorable, and gold flows out. After the fall of prices facilitates the liquidation, the trade balance becomes favorable, and gold returns.[2] Here the Ricardian analysis is *definitely* linked to *cyclical* fluctuations. The period of rising prices and gold exports is long; while the fall of prices is rapid, and the return of gold is consummated in a relatively short time.[3] And lastly, he maintains that cycles are simultaneous for England and France (with a time lag of not more than a year) and the situation is nearly the same for these two countries and the United States.[4]

Both of these theorists showed that an import of gold would raise prices immediately if spent for goods. But if advanced to entrepreneurs it would first affect interest rates, and ultimately prices, unless the stimulation given to production offset the tendency toward increased prices.[5]

It remained for Marshall to explore this process more definitely,

---

[1] "Money continued to flow to Egypt and India, but the cash which came from the provinces sufficed to make these remittances." *Ibid.*, p. 78.

"Gold flows back from the provinces to the Bank of England, since there is only an insignificant diminution of the metallic reserve, in spite of great exportations to foreign countries." *Ibid.*, p. 80. In the last quotation use is made of the device employed later in the present study to determine the flow to the interior: namely, the difference in the metallic stock of the Bank not due to foreign movements of specie must be accounted for by internal flows.

[2] Clement Juglar, *Des crises commerciales et de leur retour périodique en France, en Angleterre, et aux Etats-Unis*, Paris, 1862, p. 20, and pp. 192–199. I have used here the second edition, Paris, 1889. See also Juglar, *Du change et de la liberté d'émission*, Paris, 1868, pp. 469–470.

[3] *Ibid.*, pp. 175, 176.

[4] *Ibid.*, p. 176, and *Des crises commerciales et de leur retour périodique en France, en Angleterre, et aux Etats-Unis*, p. 255.

[5] See Angell, *The Theory of International Prices*, pp. 276–277.

and to show its relationship to internal currency flows.[1] In his view a new supply of gold would first increase bank reserves. Lower discount rates would result, and more speculation would be indulged in.[2] As a consequence prices would rise, and then people would need more hand-to-hand circulation. The new currency drawn into circulation would support and sustain the rise of prices. Marshall qualifies the analysis by showing that the changes outlined above will occur only if "other things" remain the same.[3] He examines some of these other factors, and finds that at times they will outweigh the monetary elements, and that the expected price change will not result. The other important factors are the methods of doing business — particularly the customs and practices of banking — and the amount of currency customarily carried by the people.

[1] *Official Papers by Alfred Marshall*, edited by J. M. Keynes, London, 1926. Angell shows that the analysis was first developed by Sidgwick and Giffin. *Op. cit.*, pp. 117–121 and 123.

[2] As Angell points out, speculation means all types of operations in Lombard Street, and not merely stock exchange speculation. *Ibid.*, p. 123.

[3] *Op. cit.*, pp. 38–45.

CHAPTER III

GOLD AND CURRENCY MOVEMENTS

(1) During the years from 1881 to the outbreak of the war in 1914, England [1] affords a peculiarly interesting and instructive field for the investigation of the influence of both international and domestic flows of specie upon banking and industrial conditions. Although there were wars in the Orient, the Balkans, and in America, and general European wars were narrowly averted on several occasions, England was at peace except for the South African war. The Boer War seriously affected industrial and financial conditions within England, and her international balance of payments was disturbed, but it was not catastrophic and there was no disruption of the monetary system or resort to forced issues of paper currency. From time to time during the period considered there were internal difficulties of an industrial nature, but these did not interrupt seriously the marked expansion of trade and production which occurred during these years.

On the financial side England was in a wholesome condition. Government finances were maintained in a satisfactory state, and under the administration of Lord Goschen an appreciable reduction of the interest on the outstanding Consol debt was achieved. Moreover, by the beginning of the period the Bank of England, following the oft-times repeated criticism of Bagehot and others, had begun to operate in a systematic and well-defined manner. While its control over the London money market was weakened by the large accumulation of liquid capital in the hands of the constituents of the outside market, it exerted a strong psychological influence upon this group and a certain amount of mechanical control over the business of the country outside of London. Its con-

[1] This study is confined to England and Wales, and both, for sake of brevity, will be referred to as England. The Scotch and Irish banks operated under different laws and are not considered, except in so far as they withdrew gold from London to cover excess issues of notes. In this respect the drain to these two countries resembled that to the interior of England, and is considered in connection with the latter. See below, p. 60.

trol of the local market was complete at times, and it succeeded in developing auxiliary weapons to aid its major device of the discount rate. The Bank was successful in preventing panics, and the crises of the period were far less severe than those experienced earlier in the nineteenth century.

All this was achieved in spite of the constant pressure brought upon the London bullion market — the only free market of any size — during the last two decades of the last century by the increasing relative scarcity of gold in the world, and in the years immediately preceding the Great War by the demands of the Continental nations for larger specie reserves. On the other hand, the amalgamation movement in English banking proceeded rapidly during the whole period, and the banking resources of the nation were gradually concentrated in the hands of a relatively small number of companies. While there was not always the closest sympathy between the banks and the Bank of England, relations were improved toward the end of the period, and evidences of co-operation were seen at times.

(2) In addition to an important secular growth of industry and commerce during the period, there were pronounced cyclical fluctuations in business and prices. For the present purpose it is unnecessary to distinguish between all of the phases of the trade cycles, since variations in the volume of currency and credit and movements of specie would tend to be alike in the stages of revival and prosperity and during recession and depression. The analysis of Hawtrey assumes that gold will flow from the central reserve into circulation during the period of expansion,[1] and specie may be exported if foreign countries are expanding at a slower rate. While the movement of gold may lag somewhat behind the increased volume of business and the rise of prices, losses of specie from the central bank and a reduction in the reserve ratio would accompany the revival and prosperity stages of the business cycle. In periods of recession and depression, reserves would rise as

---

[1] Under a proportional reserve system of note issue, the reserve ratio would be affected, but no decline in actual reserves would occur, providing there were no gold coins or certificates in circulation, and no gold was withdrawn for export.

coin and notes flowed in from the circulation, and specie imports would occur if the contraction of credit was less severe in foreign countries.

The fluctuations of business and prices attributable to cyclical or accidental forces are classified below into periods of expansion and recession:

I.	Expansion	1st quarter, 1881 — 4th quarter, 1882
	Recession	1st quarter, 1883 — 2nd quarter, 1886
II.	Expansion	3rd quarter, 1886 — 3rd quarter, 1890
	Recession	4th quarter, 1890 — 1st quarter, 1895
III.	Expansion	2nd quarter, 1895 — 1st quarter, 1900
	Recession	2nd quarter, 1900 — 3rd quarter, 1904
IV.	Expansion	4th quarter, 1904 — 2nd quarter, 1907
	Recession	3rd quarter, 1907 — 4th quarter, 1908
V.	Expansion	1st quarter, 1909 — 4th quarter, 1912
	Recession	1st quarter, 1913 — 2nd quarter, 1914

The above classification is made from the quarterly indexes of business conditions and prices prepared by Persons, Silberling, and Berridge [1] for the period 1903 to 1914 and by Dr. Thomas [2] for the earlier years. The indexes are shown in Chart I and Table I. While the division of business fluctuations into periods of expansion and recession is not precisely accurate so far as the turning points are concerned, the general course of the business cycles is fairly well defined.[3]

[1] Warren M. Persons, Norman J. Silberling, and William A. Berridge, "An Index of British Economic Conditions, 1903–1914," *Review of Economic Statistics*, Preliminary vol. IV, Supplement no. 2 (June, 1922).

[2] Dorothy Swaine Thomas, "An Index of British Business Cycles," *Journal of the American Statistical Association*, New Series, vol. XXI, no. 153 (March, 1926), pp. 60–63. This index became more comprehensive as new series were added to its composition in later years.

[3] The data collected by Willard L. Thorp in *Business Annals*, New York, 1926, affords a comprehensive, though terse, description of the major movements of each year, and is useful as an indication of contemporary opinion. While the latter may not be an altogether sound criterion of cycles, it is of particular significance for gold movements, as the latter are peculiarly sensitive to psychological influences acting upon the money market. For an admirable and judicious summary of the usefulness of this method of treatment and its relation to statistical indices, see Wesley C. Mitchell, *Business Cycles; The Problem and Its Setting*, New York, 1928, pp. 365–366. An index of the value of exports is used by Dr. A. G. Silverman as an approximate description of the business cycle. It corresponds closely with the Thomas index except in 1896 and 1897, when export values lie above the trend, in agreement with Thorp's description. A somewhat longer period of depression is shown from 1902 to 1905, and the prosperity is shown more clearly during 1913. See Silverman, *op. cit.*, p. 121.

CHART I

QUARTERLY INDEXES OF BRITISH BUSINESS CYCLES, 1881–1914

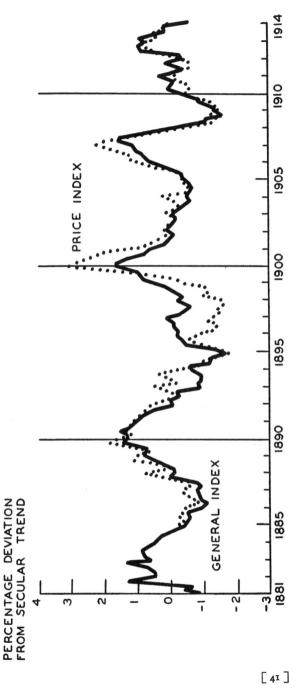

TABLE I

QUARTERLY INDEXES OF BRITISH BUSINESS CYCLES, 1881–1914 *

(*Percentage Deviations of Actual Items — with seasonal variation allowed for — from Secular Trend, in terms of Standard Deviation*)

Year	General Index	Sauerbeck's Price Index	Year	General Index	Sauerbeck's Price Index
1881	−1.10		1891	+1.05	+1.06
	− .34			+ .98	+ .90
	− .62			+ .71	+ .76
	+1.36			+ .55	+ .63
1882	+ .50		1892	− .09	+ .39
	+ .50			− .03	+ .23
	+ .72			− .23	− .25
	+1.43			− .11	+ .14
1883	+ .66		1893	− .91	+ .47
	+ .85			− .78	− .09
	+ .88			− .96	+ .05
	+ .83			− .95	+ .44
1884	+ .48		1894	− .65	− .37
	+ .34			− .66	− .78
	+ .21			−1.20	−1.01
	− .07			−1.30	−1.47
1885	− .37	− .23	1895	−1.65	−1.79
	− .49	− .31		−1.26	−1.22
	− .56	− .25		− .43	− .68
	− .46	− .31		− .31	− .58
1886	− .92	− .47	1896	− .15	− .86
	−1.04	− .87		− .21	−1.27
	− .90	− .65		− .04	−1.40
	− .86	− .33		− .02	−1.07
1887	− .74	− .33	1897	+ .13	−1.26
	− .91	− .58		− .27	−1.49
	− .67	− .54		− .45	−1.34
	+ .01	+ .23		− .62	−1.59
1888	− .11	+ .76	1898	− .22	−1.48
	− .07	− .09		− .30	− .98
	+ .18	+ .31		− .22	−1.09
	+ .43	+1.11		+ .18	−1.03
1889	+ .83	+1.07	1899	+ .53	− .41
	+ .79	+ .59		+ .79	− .03
	+ .77	+ .97		+ .84	+ .75
	+1.51	+1.82		+1.19	+2.05
1890	+1.36	+1.60	1900	+1.71	+3.07
	+1.33	+1.22		+1.64	+2.62
	+1.52	+1.44		+1.29	+2.52
	+1.26	+1.28		+ .75	+1.91

* SOURCE: D. S. Thomas, "An Index of British Business Cycles," *Journal of the American Statistical Association*, New Series, vol. XXI, no. 153 (March, 1926), Table II, facing p. 61.

TABLE I (*Continued*)

Year	General Index	Sauerbeck's Price Index	Year	General Index	Sauerbeck's Price Index
1901	+ .59	+1.01	1908	+ .02	− .64
	+ .20	+ .61		− .80	−1.28
	+ .05	+ .45		−1.19	−1.40
	− .04	+ .17		−1.63	−1.24
1902	+ .06	+ .20	1909	−1.28	−1.49
	.00	+ .17		−1.31	−1.34
	− .03	− .03		− .93	−1.12
	− .02	− .12		− .80	− .71
1903	− .02	+ .16	1910	− .58	− .50
	− .14	− .08		+ .03	− .61
	− .31	− .25		− .10	− .56
	− .60	− .14		− .11	− .19
1904	− .48	+ .27	1911	+ .38	− .09
	− .58	− .45		− .02	− .08
	− .68	− .56		− .42	− .61
	− .56	− .37		+ .15	− .68
1905	− .46	− .47	1912	− .45	− .36
	− .35	− .42		− .26	+ .11
	− .07	− .03		+ .91	+ .50
	+ .32	+ .59		+ .98	+ .81
1906	+ .68	+ .81	1913	+ .74	+ .73
	+ .75	+1.23		+ .97	+ .48
	+ .79	+1.24		+ .49	+ .48
	+1.01	+1.90		+ .12	+ .25
1907	+1.15	+2.11	1914	+ .06	− .16
	+1.53	+2.15		− .62	− .65
	+1.41	+1.35			
	+ .75	+ .28			

In the following sections the cyclical fluctuations of business are examined in relation to the international flow of gold and the domestic movement of currency, and in the succeeding chapter variations in bank credit and discount rates are considered with respect to "secondary" expansion. In later chapters a more minute historical survey is made of the developments in the London money market and the banking system.

(3) The London gold market was preeminently important during this period. A very large amount of specie moved direct to England from the mining countries, especially South Africa and Australia, to be sold on the bullion market. Before the War most

of the supplies from these regions were sold in London.[1] The presence of this market was of great value to England, for it gave her immediate access to newly arriving supplies. Not only was it profitable to the dealers in the way of commissions but a much slighter change in discount rates reacting on exchange rates was necessary to retain gold already within the country than to cause it to flow in from abroad. And the fact that at all times shipments were on the seas bound for London from the mines gave security to her reserve position.[2]

It was the practice for newly mined gold to be sent to the London refiners, and then to be sold through the various firms of bullion dealers. The price never fell below the Bank of England buying rate of 77s. 9d., but it often went considerably above it. With the exception of certain periods, whenever the slightest fraction above the Bank price was offered by agents acting for principals in foreign countries, the gold was sold to them for export. At times, however, the Bank came into the market as an active bidder for specie at a price above its usual buying rate.[3] Instances of this practice will be noticed later, but for the present it is desirable merely to point out the limits to which this policy might go. Rarely did the Bank pay as much as 77s. 10d., and 77s. 10½d. was the upper limit.[4] In the case of sales of gold by the Bank, the price of bullion was normally at 77s. 10½d. per ounce, but it might rise above that to the limit of the tolerance of the Mint. Beyond 77s. 11d. it was not safe to go, as coin would then be de-

---

[1] In the post-war period considerable amounts of gold were shipped directly to other countries. See Paul Einzig, *International Gold Movements*, London, 1929, pp. 15–17, and William A. Brown, Jr., *England and the New Gold Standard, 1919–1926*, New Haven, 1929, pp. 136–137, 185–187, and 281–283.

[2] See, e.g., J. Herbert Tritton, "The Short Loan Fund of the London Money Market," *Journal of the Institute of Bankers*, vol. XXIII (March, 1902), p. 104, and Brown, *op. cit.*, pp. 284–285.

[3] See A. C. Whitaker, *Foreign Exchange*, New York, 1920, Ch. XX, and *Interviews on the Banking and Currency Systems of England, Scotland, France, Germany, Switzerland, and Italy*, National Monetary Commission, 1910, Senate Document No. 405, 61st Congress, 2nd Session, pp. 29 and 103. See also *The Economist*, London, and the *Bankers' Magazine*, London, *passim*. The latter was titled the *Bankers' Magazine, Journal of the Money Market, and Commercial Digest* from 1854 to 1889, when its title was changed to the *Bankers' Insurance Managers' and Agents' Magazine*. It is referred to hereafter as the *Bankers' Magazine*.

[4] Whitaker, *op. cit.*, p. 504, and *Interviews, op. cit.*, p. 103.

manded to be melted down for export.[1] The buying and selling prices of *foreign coins* were subject to somewhat greater variation, however, depending upon the needs of the Bank.[2]

Statistics relating to international gold movements for England were published monthly by the Board of Trade, and movements at London weekly by the London Customs House. In addition the Bank of England reported daily the amount of bullion and coin received and withdrawn on foreign account. Specie which entered England only to be reshipped immediately after sale in the bullion market, and movements into bank reserves or the internal circulation which were offset shortly by exports from the same sources, had very little effect upon the international balance of payments.[3] And movements which were cancelled by others in the opposite direction within a week could have little effect even upon the fluctuations of commodity prices and industrial production. Accordingly, only the net weekly figures of the Bank of England [4] and the London Customs House have been utilized here.

In order to secure series more nearly comparable with the index of business cycles, quarterly totals of the weekly net figures have been computed, and they are shown in Table II and Chart II. Four-quarter moving averages of each series have been calculated,[5] and they are shown in Table III and Chart III. Random fluctuations have been reduced by the moving average, but seasonal variation and secular trend are evident. It has not been thought worth while to remove the seasonal and the trend in the usual manner partly because the cyclical variations are very

[1] Whitaker, *op. cit.*, pp. 558–560, and George Clare, *A Money-Market Primer, and Key to the Exchanges*, London, 1891, p. 22.

[2] See *Bankers' Magazine, passim.*

[3] See, however, White, *op. cit.*, pp. 30–32. Important effects upon the money market can be exerted by gold flows during the early part of a quarter, even though they are balanced by exports during the latter part. The weekly figures are considered in the historical sections below, where it seemed important, although quarterly figures are used for the determination of cyclical fluctuations.

[4] The daily reports were compiled into a weekly series by the *Economist*, and the convenient yearly summaries published in its annual Commercial History and Review have been used as the source for this series.

[5] In calculating the series, the movements for weeks which fell into two quarters were arbitrarily assigned to the quarter containing four or more days of the week.

TABLE II

Quarterly Net Foreign Movements of Gold, England, 1881–1913 *

(In thousands of pounds)

Quarter Ending †	Reported by Bank of England	Reported by London Customs House	Quarter Ending	Reported by Bank of England	Reported by London Customs House
Mar. 31, 1881..	+ 1,370	− 515	Dec. 31.......	+ 5,159	+ 5,081
June 30.......	+ 369	− 168	Mar. 31, 1891..	− 3,800	− 2,646
Sept. 30.......	− 2,612	− 2,669	June 30.......	+ 6,777	+10,113
Dec. 31.......	− 1,467	− 2,256	Sept. 30.......	− 2,743	− 677
Mar. 31, 1882..	+ 1,304	+ 2,675	Dec. 31.......	− 1,792	− 727
June 30.......	+ 1,730	+ 2,232	Mar. 31, 1892..	− 1,247	− 290
Sept. 30.......	− 654	− 1,225	June 30.......	+ 2,566	+ 3,025
Dec. 31.......	+ 61	+ 301	Sept. 30.......	+ 1,782	+ 3,490
Mar. 31, 1883..	− 485	+ 122	Dec. 31.......	− 3,039	+ 443
June 30.......	+ 238	− 369	Mar. 31, 1893..	− 939	+ 407
Sept. 30.......	+ 2,721	+ 2,153	June 30.......	+ 4,992	+ 5,766
Dec. 31.......	− 1,672	− 1,771	Sept. 30.......	− 2,489	− 60
Mar. 31, 1884..	+ 2,300	+ 1,576	Dec. 31.......	− 1,590	− 1,222
June 30.......	+ 864	+ 1,163	Mar. 31, 1894..	+ 2,727	+ 2,906
Sept. 30.......	− 1,270	− 1,790	June 30.......	+ 9,034	+ 9,808
Dec. 31.......	− 930	− 1,299	Sept. 30.......	+ 506	+ 2,935
Mar. 31, 1885..	+ 2,784	+ 1,496	Dec. 31.......	− 3,518	− 3,502
June 30.......	+ 2,002	+ 2,486	Mar. 31, 1895..	+ 2,331	+ 1,892
Sept. 30.......	− 1,730	− 1,991	June 30.......	+ 1,699	+ 2,161
Dec. 31.......	− 1,997	+ 46	Sept. 30.......	+ 7,131	+ 7,533
Mar. 31, 1886..	− 1,156	+ 1,508	Dec. 31.......	+ 3,215	+ 3,335
June 30.......	− 596	− 1,006	Mar. 31, 1896..	+ 253	+ 1,447
Sept. 30.......	+ 293	+ 738	June 30.......	+ 2,600	+ 2,794
Dec. 31.......	− 2,521	− 2,008	Sept. 30.......	− 6,118	− 4,714
Mar. 31, 1887..	+ 2,584	+ 1,868	Dec. 31.......	− 4,834	− 5,134
June 30.......	− 323	+ 28	Mar. 31, 1897..	+ 2,244	+ 1,326
Sept. 30.......	− 1,322	− 1,683	June 30.......	− 174	+ 1,508
Dec. 31.......	− 135	+ 502	Sept. 30.......	− 1,880	− 724
Mar. 31, 1888..	+ 690	+ 1,364	Dec. 31.......	− 1,250	− 2,324
June 30.......	− 315	+ 372	Mar. 31, 1898..	− 1,483	− 23
Sept. 30.......	− 468	− 356	June 30.......	+ 7,251	+ 8,904
Dec. 31.......	+ 126	− 800	Sept. 30.......	− 2,868	− 2,256
Mar. 31, 1889..	+ 1,092	+ 1,298	Dec. 31.......	− 953	+ 905
June 30.......	+ 3,260	+ 3,749	Mar. 31, 1899..	− 216	+ 153
Sept. 30.......	− 1,800	− 574	June 30.......	+ 2,203	+ 2,955
Dec. 31.......	− 1,194	− 1,818	Sept. 30.......	+ 5,985	+ 8,159
Mar. 31, 1890..	+ 3,324	+ 2,888	Dec. 31.......	+ 1,519	− 852
June 30.......	− 1,101	+ 304	Mar. 31, 1900..	+ 795	+ 4,894
Sept. 30.......	+ 1,414	+ 969	June 30.......	+ 881	+ 537

* Source: The annual Commercial History and Review of the *Economist*.
† The figures for weeks which overlapped two quarters were allotted to the quarter which included four or more days.

TABLE II (*Continued*)

Quarter Ending	Reported by		Quarter Ending	Reported by	
	Bank of England	London Customs House		Bank of England	London Customs House
Sept. 30	+ 3,602	+ 5,501	June 30	+ 2,826	+ 2,032
Dec. 31	− 4,280	− 2,328	Sept. 30	+ 2,799	+ 4,316
Mar. 31, 1901	+ 2,387	+ 2,703	Dec. 31	− 975	− 3,779
June 30	+ 3,072	+ 4,506	Mar. 31, 1908	+ 2,543	+ 4,632
Sept. 30	+ 1,238	+ 2,563	June 30	+ 2	− 857
Dec. 31	− 3,627	− 3,454	Sept. 30	− 1,862	− 2,840
Mar. 31, 1902	+ 324	+ 1,429	Dec. 31	− 3,692	− 4,618
June 30	+ 4,193	+ 4,756	Mar. 31, 1909	+ 7,801	+ 5,927
Sept. 30	− 1,308	+ 1,020	June 30	+ 2,387	+ 3,554
Dec. 31	− 3,455	− 894	Sept. 30	− 2,212	− 1,865
Mar. 31, 1903	+ 1,732	+ 2,963	Dec. 31	+ 143	+ 227
June 30	+ 3,097	+ 3,041	Mar. 31, 1910	− 2,937	− 1,134
Sept. 30	− 2,291	− 1,917	June 30	+11,391	+11,611
Dec. 31	− 2,835	− 3,131	Sept. 30	− 4,151	− 2,816
Mar. 31, 1904	+ 1,510	+ 1,386	Dec. 31	− 341	− 547
June 30	+ 1,845	+ 1,561	Mar. 31, 1911	+ 3,101	+ 2,855
Sept. 30	+ 4,030	+ 3,982	June 30	+ 5,641	+ 5,111
Dec. 31	− 4,569	− 5,766	Sept. 30	+ 2,924	+ 3,254
Mar. 31, 1905	+ 4,204	+ 5,308	Dec. 31	− 2,580	− 2,604
June 30	+ 2,225	+ 2,704	Mar. 31, 1912	+ 194	+ 1,092
Sept. 30	− 3,246	− 594	June 30	+ 5,245	+ 4,388
Dec. 31	− 2,694	+ 895	Sept. 30	− 10	+ 2,364
Mar. 31, 1906	+ 5,037	+ 5,562	Dec. 31	− 2,835	− 2,609
June 30	+ 1,415	− 794	Mar. 31, 1913	+ 2,066	+ 1,539
Sept. 30	− 1,828	− 749	June 30	+ 5,567	+ 6,189
Dec. 31	+ 1,138	− 809	Sept. 30	+ 2,651	+ 6,458
Mar. 31, 1907	+ 1,643	+ 3,075	Dec. 31	+ 3,563	− 214

marked after smoothing, and partly because the series of internal movements of currency [1] with which external movements of gold are to be compared are subject to considerable error and removal of seasonal fluctuations and trend seems unwarranted. The resemblance between the curve of gold movements reported by the London Customs House on Chart III and the curve of net gold movements shown by Dr. Silverman [2] is close. The latter curve shows a series of monthly gold movements as reported by the Board of Trade after seasonal variation has been removed. The curve derived from the Bank of England returns differs from

[1] See below, p. 55, for the method of deriving these series.
[2] Silverman, *op. cit.*, pp. 120–121. See also p. 115.

CHART II

Net Quarterly External Movements of Gold and Internal Movements of Currency, England, 1881–1913

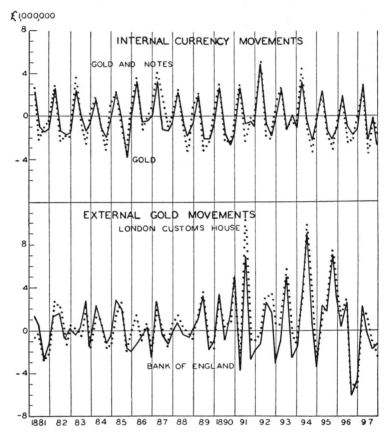

CHART II

(*Continued*)

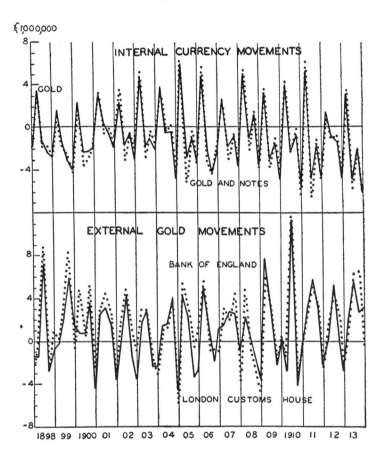

that derived from the London Customs House data, especially from 1891 to 1894, from 1898 to 1903, and in 1905. The cyclical fluctuations in the former series are in general more pronounced, as is to be expected. The more important differences between the series are considered below.[1]

(4) Internal movements of gold and currency also reveal pronounced cyclical variation. England continued to use gold sovereigns and half-sovereigns for the greater part of her internal circulation until the war, but substantial amounts of silver and bronze coins and Bank of England and country bank notes were also or use. None of the bank notes were less than £5 in denomination, however, and they were but little used in ordinary transactions. Moreover, it was necessary to endorse them in making payments, and this bother helped to make checking accounts more desirable. The requirements of the Act of 1844 prevented any large fluctuation in the Fiduciary issue of Bank of England notes. For the most part notes were used, consequently, only in payment of large salaries, for traveling expenses, and as till money for banks.[2] The latter use was especially important, for country banks could retain their own notes instead of metal money for reserve purposes beyond the daily coin requirements. The country bank note issues steadily declined during the period, as these banks were required to relinquish the privilege of issue upon bankruptcy or upon amalgamation with a bank maintaining

---

[1] See pp. 64 ff., and 124.

[2] See F. Straker, *The Money Market*, London, 1904, p. 60. The advantage of note issue to the country banks was noted by the *Bankers' Magazine*: "But besides the amount in circulation, the possession of the power of issue gives to a bank which has it another advantage — the use of the 'till money' which the unissued notes supplies. At many of the branches of the 125 banks referred to above [note-issuing banks] a very considerable part of the money kept to meet immediate requirements is in the form of notes of the bank itself, which, for this purpose, are as serviceable as coin or notes of the Bank of England." June, 1889, pp. 670–671. Cf. also George Clare, *op. cit.*, p. 19, and F. Lavington, *The English Capital Market*, London, 1921, second edition, 1929, p. 38.

An increase of some 5 to 7 million in the "circulation" of Bank of England notes was believed to have resulted from the increased number of banking offices (some 1,700 in twenty years). The larger part of this increase in the holding of Bank notes in the tills of other banks came between 1890 and 1895. See *Bankers' Magazine*, August, 1896, p. 139.

CHART III

FOUR-QUARTER MOVING AVERAGES OF NET EXTERNAL GOLD MOVEMENTS, ENGLAND, 1881–1913

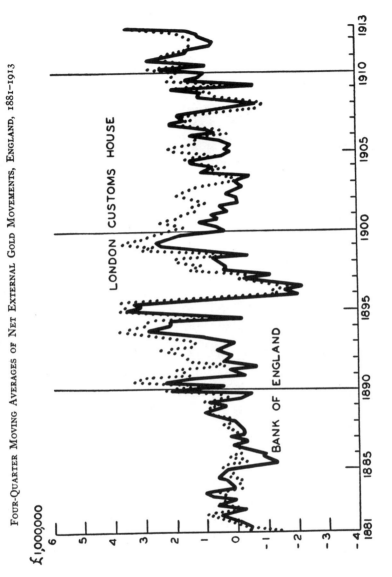

TABLE III

Four-Quarter Moving Averages of Net Foreign Movements of Gold, England, 1881–1913 *

(In thousands of pounds)

Centered at †	From reports by Bank of England	From reports by London Customs House	Centered at	From reports by Bank of England	From reports by London Customs House
July 1, 1881 ...	− 585	−1,402	Jan. 1, 1891 ...	+2,388	+3,379
Oct. 1	− 602	− 604	April 1.........	+1,348	+2,968
Jan. 1, 1882 ...	− 261	− 7	July 1.........	− 390	+1,516
April 1.........	+ 226	+ 354	Oct. 1.........	+ 249	+2,105
July 1.........	+ 610	+ 996	Jan. 1, 1892 ...	− 804	+ 333
Oct. 1.........	+ 163	+ 358	April 1.........	+ 327	+1,400
Jan. 1, 1883 ...	− 210	− 293	July 1.........	+ 16	+1,667
April 1.........	+ 634	+ 552	Oct. 1.........	+ 92	+1,841
July 1.........	+ 200	+ 34	Jan. 1, 1893 ...	+ 699	+2,526
Oct. 1.........	+ 897	+ 397	April 1.........	+ 369	+1,389
Jan. 1, 1884 ...	+1,053	+ 778	July 1.........	− 6	+1,223
April 1.........	+ 56	− 206	Oct. 1.........	+ 910	+1,848
July 1.........	+ 241	− 88	Jan. 1, 1894 ...	+1,920	+2,858
Oct. 1.........	+ 362	− 108	April 1.........	+2,669	+3,807
Jan. 1, 1885....	+ 646	+ 223	July 1.........	+2,187	+3,037
April 1.........	+ 532	+ 173	Oct. 1.........	+2,088	+2,783
July 1.........	+ 265	+ 509	Jan. 1, 1895 ...	+ 254	+ 872
Oct. 1.........	− 720	+ 512	April 1.........	+1,911	+2,021
Jan. 1, 1886 ...	−1,370	− 361	July 1.........	+3,594	+3,730
April 1.........	− 864	+ 322	Oct. 1.........	+3,074	+3,619
July 1.........	− 995	− 192	Jan. 1, 1896 ...	+3,300	+3,777
Oct. 1.........	− 60	− 102	April 1.........	− 12	+ 716
Jan. 1, 1887 ...	+ 8	+ 158	July 1.........	−2,025	−1,402
April 1.........	− 396	− 449	Oct. 1.........	−1,527	−1,182
July 1.........	+ 201	+ 179	Jan. 1, 1897 ...	−2,220	−1,754
Oct. 1.........	− 272	+ 53	April 1.........	−1,161	− 756
Jan. 1, 1888 ...	− 270	+ 139	July 1.........	− 265	− 56
April 1.........	− 57	+ 470	Oct. 1.........	−1,197	− 381
July 1.........	+ 8	+ 145	Jan. 1, 1898 ...	+ 610	+1,458
Oct. 1.........	+ 109	+ 128	April 1.........	+ 412	+1,075
Jan. 1, 1889 ...	+1,002	+ 973	July 1.........	+ 487	+1,882
April 1.........	+ 670	+ 918	Oct. 1.........	+ 804	+1,926
July 1.........	+ 340	+ 664	Jan. 1, 1899 ...	− 458	+ 439
Oct. 1.........	+ 898	+1,061	April 1.........	+1,755	+3,043
Jan. 1, 1890 ...	− 193	+ 200	July 1.........	+2,373	+2,604
April 1.........	− 611	+ 586	Oct. 1.........	+2,626	+3,789
July 1.........	+2,199	+2,310	Jan. 1, 1900 ...	+2,295	+3,184
Oct. 1.........	+ 418	+ 927	April 1.........	+1,699	+2,520

* Source: Table II.

† The center lies approximately on the first day of each quarter. Since the figures for overlapping weeks were allotted to the quarter which included four or more days, the center varies within a limit of three days on either side of the first day of a quarter.

TABLE III (*Continued*)

Centered at	From reports by Bank of England	From reports by London Customs House	Centered at	From reports by Bank of England	From reports by London Customs House
July 1.........	+ 250	+2,151	April 1.........	+2,102	+2,156
Oct. 1.........	+ 648	+1,603	July 1.........	+1,573	+1,411
Jan. 1, 1901 ...	+1,195	+2,596	Oct. 1.........	+1,798	+1,800
April 1.........	+ 604	+1,861	Jan. 1, 1908 ...	+1,092	+1,078
July 1.........	+ 768	+1,580	April 1.........	− 73	− 711
Oct. 1.........	+ 252	+1,261	July 1.........	− 752	− 921
Jan. 1, 1902 ...	+ 532	+1,324	Oct. 1.........	+ 562	− 597
April 1.........	− 104	+ 938	Jan. 1, 1909 ...	+1,158	+ 506
July 1.........	− 62	+1,578	April 1.........	+1,071	+ 749
Oct. 1.........	+ 290	+1,961	July 1.........	+2,030	+1,961
Jan. 1, 1903 ...	+ 16	+1,532	Oct. 1.........	− 655	+ 196
April 1.........	− 229	+ 798	Jan. 1, 1910 ...	+1,596	+2,210
July 1.........	− 74	+ 234	April 1.........	+1,112	+1,972
Oct. 1.........	− 130	− 155	July 1.........	+ 990	+1,778
Jan. 1, 1904 ...	− 443	+ 525	Oct. 1.........	+2,500	+2,776
April 1.........	+1,138	+ 949	Jan. 1, 1911 ...	+ 812	+1,151
July 1.........	+ 704	+ 291	April 1.........	+2,831	+2,668
Oct. 1.........	+1,378	+1,271	July 1.........	+2,271	+2,154
Jan. 1, 1905 ...	+1,472	+1,557	Oct. 1.........	+1,545	+1,713
April 1.........	+ 346	+ 413	Jan. 1, 1912 ...	+1,446	+1,532
July 1.........	+ 122	+2,078	April 1.........	+ 712	+1,310
Oct. 1.........	+ 330	+2,142	July 1.........	+ 648	+1,309
Jan. 1, 1906 ...	+ 128	+1,268	Oct. 1.........	+1,116	+1,420
April 1.........	+ 482	+1,228	Jan. 1, 1913 ...	+1,197	+1,871
July 1.........	+1,440	+ 804	April 1.........	+1,862	+2,894
Oct. 1.........	+ 592	+ 156	July 1.........	+3,462	+3,493
Jan. 1, 1907....	+ 945	+ 888			

an office within sixty-five miles of London.[1] There was a steady diminution of these notes, and the growth of the Bank of England issue was accounted for largely by the replacement of the country bank notes. The Bank was permitted to increase its issue to the extent of two-thirds of the relinquished issue.[2]

Estimates of the amount of gold and silver coin in circulation

[1] Acts. *7th* and *8th Vict.*, *cap.* 32 (1844).

[2] The country note issue outstanding in the hands of the public on January 8, 1881, was £3,497,000, while it had declined to only £118,000 on December 27, 1913. On January 5, 1881, the Bank of England circulation outstanding amounted to £26,954,000, and on December 31, 1913, it was £29,608,000. See the London Gazette. Cf., however, F. H. Jackson, "To What Extent Has the Position of the Bank of England Changed in Recent Years in Comparison with (a) Similar Insti-

in England were made from time to time, but they are of little use in the determination of cyclical fluctuation of the currency.[1] The annual statistics prepared by the Mint each year showing the

tutions Abroad, and (b) London Clearing Bankers?" an address delivered before the Political Economy Club, June 12, 1903, privately printed, London, 1903, p. 5.

The increases in the Bank of England issue are conveniently summarized in W. F. Spalding, *The London Money Market*, London, 1922, p. 45:

Bank of England Fiduciary Issue

Authorized by the Bank Charter	
Act of 1844	£14,000,000
Authorized by Order in Council	
Act of 7th December 1855	475,000
10th July 1861	175,000
21st February 1866	350,000
1st April 1881	750,000
15th September 1887	450,000
8th February 1890	250,000
29th January 1894	350,000
3rd March 1900	975,000
11th August 1902	400,000
10th August 1903	275,000
Total	£18,450,000

[1] The various estimates are gathered together in the *Annual Report of the Deputy Master and Comptroller of the Mint*, 1903, p. 21 (cited hereafter as *Mint Report*).

Estimated Gold Supply of the United Kingdom at Intervals, 1844–1903

1844   £46,000,000 (Mr. Newmarch's estimate)
1856   £75,000,000 (Mr. Newmarch's estimate)
1868 under £80,000,000 (Mr. Jevons' estimate)
1883   £110,000,000 (Mr. Inglis-Palgrave's estimate — "a wide limit")
1888   £102,500,000 (The Royal Mint's Estimate)
1892   £90,000,000 (The Chancellor of the Exchequer's estimate)
1895 In active circulation £62,500,000
       In reserves held by banks £30,000,000 (The Royal Mint's estimate)
1903   £63,500,000 in active circulation.

Beginning with 1907 the Mint collected statistics of gold holdings by banks; see below, p. 90.

In 1914 the total gold circulation was estimated at £123,000,000, of which £44,500,000 was held by banks. See *Mint Report*, 1914, p. 9. This estimate was accepted by the Cunliffe Committee in 1918. See the *First Interim Report of the Committee on Currency and Foreign Exchanges after the War*, Cmd. 9182 (1918), par. 13.

The Mint estimates were based upon various methods of computing the coins in circulation in accordance with the coins received and issued under the coinage acts of 1889 and 1891. See *Mint Report*, 1890, pp. 95–96; 1896, pp. 91–92; and 1903, p. 21.

value of coins struck and issued, and the value of coins withdrawn from circulation for recoinage are of help, however. When these figures are combined with the statistics of foreign movement of British gold coin, a series is secured which throws some light on the internal movements of coin. These figures have been plotted in Charts IV and V to show the movements of gold coin, silver coin, and bronze coin separately, and a three-year moving average has been prepared for the gold coin series to remove some of the irregularity.[1]

(5) A second method of ascertaining the cyclical variation in the internal currency involves a process of deduction from certain figures contained in the weekly return of the Bank of England. From this statement it is easy to determine the amount of specie held by the Bank. In addition to gold a small amount of silver was held by the Banking Department, but it was generally believed that the sum was small; and variations in it were relatively slight in comparison with changes in the gold holdings. The difference between the specie holdings for successive weeks indicates the amount of gold which must have moved through either international or domestic channels. The Bank returns of gold received and taken for export [2] are then combined with the figure for

The totals do not include the gold held at the Bank of England. See also R. A. Lehfeldt, *Gold, Prices, and the Witwatersrand*, London, 1919, pp. 121–122; W. A. Shaw, *Currency, Credit and the Exchanges*, London, 1927, p. 33; and *The Theory and Principles of Central Banking*, London, 1930, p. 98.

[1] See Tables IV and V. A plus sign indicates an increase in coin in circulation.

[2] The movements are complicated somewhat in certain years by the "earmarking" and release of gold on account of India. These took place in 1899–1900, and more importantly from 1905–1914.
See G. F. Shirras, *Indian Finance and Banking*, London, 1920, pp. 186, 463, and 465, and J. M. Keynes, *Indian Currency and Finance*, London, 1913, revised edition, 1924, p. 49.
The manner of treating this gold at the Bank was explicitly stated by the Governor in answer to a question at the meeting of the shareholders on September 20, 1900: "In reply to his inquiry about the India gold, it was locked up here like so much merchandise and formed no part of our account, but when released it came into our possession and formed part of the reserve." *Bankers' Magazine*, October, 1900, p. 517.
The *Economist* has taken proper account of this movement of gold in making up its figures of imports and exports in nearly all cases. And no significant changes would result in the series from the correction of minor omissions, as they were small in magnitude.

TABLE IV

MOVEMENTS OF GOLD COIN INTO AND OUT OF CIRCULATION IN ENGLAND, 1881–1913 *

(In thousands of pounds)

Year	Net Gold Coin Issued by the Mint	Net British Gold Coin Imported or Exported	Total Net Gain or Loss of Gold Coin in Circulation	Three-Year Moving Average of Total Net
1881	..	− 73	− 73	..
1882	..	+ 1,000	+ 1,000	− 363
1883	− 2,042	+ 27	− 2,015	−1,045
1884	+ 1,351	− 3,471	− 2,120	− 728
1885	+ 1,724	+ 226	+ 1,950	− 354
1886	..	− 893	− 893	+ 907
1887	− 391	+ 2,056	+ 1,665	− 644
1888	+ 365	− 3,069	− 2,704	+ 660
1889	+ 6,897	− 3,879	+ 3,018	+1,925
1890	+ 4,573	+ 987	+ 5,460	+6,147
1891	+ 6,289	+ 3,675	+ 9,964	+4,532
1892	− 3,460	+ 1,661	− 1,799	+4,456
1893	+ 2,966	+ 2,237	+ 5,203	+3,676
1894	+ 2,078	+ 5,545	+ 7,623	+3,465
1895	+ 1,411	− 3,842	− 2,431	+3,117
1896	+ 2,214	+ 1,946	+ 4,160	+2,370
1897	− 971	+ 6,351	+ 5,380	+5,475
1898	+ 3,729	+ 3,155	+ 6,884	+4,358
1899	+ 7,211	− 6,402	+ 809	+6,402
1900	+11,304	+ 209	+11,513	+4,196
1901	+ 799	− 528	+ 271	+3,353
1902	+ 4,808	− 3,234	+ 1,574	+1,593
1903	+ 8,044	− 5,110	+ 2,934	+2,658
1904	+ 8,942	− 5,477	+ 3,465	+2,339
1905	+ 3,800	− 3,181	+ 619	+2,046
1906	+ 9,465	− 7,411	+ 2,054	+3,473
1907	+18,251	−10,506	+ 7,745	+4,386
1908	+11,600	− 8,242	+ 3,358	+3,167
1909	+10,800	−12,401	− 1,601	−4,536
1910	+ 2,600	−17,966	−15,366	− 292
1911	+30,783	−14,691	+16,092	+3,583
1912	+30,103	−20,079	+10,024	+7,036
1913	+24,739	−29,747	− 5,008	..

* SOURCE: *Statistical Abstract for the United Kingdom.*

CHART IV

ANNUAL MOVEMENTS OF GOLD COIN IN CIRCULATION IN ENGLAND, 1881–1913

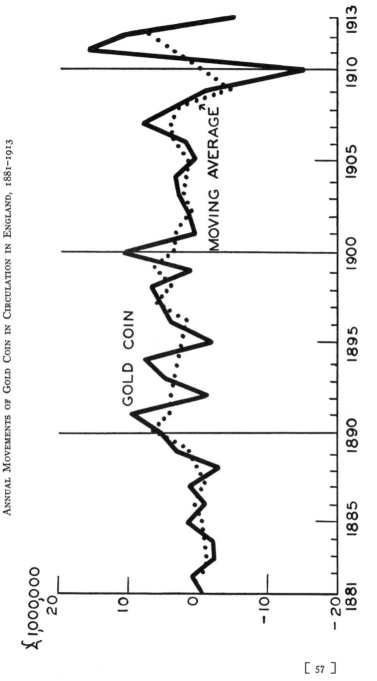

TABLE V

Net Movement of Coin Into and Out of Circulation, 1881–1913 *

(In thousands of pounds)

Year	Net Gold Coin Movement	Net Silver Coin Movement	Net Bronze Coin Movement	Total Net Movement
1881	− 73	+ 18	+ 23	− 32
1882	+ 1,000	+ 41	+ 40	+ 999
1883	− 2,015	+ 540	+ 50	− 1,425
1884	− 2,120	+ 268	+ 65	− 1,787
1885	+ 1,950	+ 83	+ 57	+ 2,090
1886	− 893	+ 197	+ 43	− 653
1887	+ 1,665	+ 397	+ 58	+ 2,102
1888	− 2,704	+ 292	+ 41	− 2,371
1889	+ 3,018	+1,566	+ 67	+ 4,651
1890	+ 5,460	+ 975	+ 90	+ 6,525
1891	+ 9,964	+ 530	+ 90	+10,584
1892	− 1,799	+ 587	+ 59	− 1,153
1893	+ 5,203	+ 534	+ 47	+ 5,784
1894	+ 7,623	+ 475	+ 33	+ 8,131
1895	− 2,431	+ 509	+ 41	− 1,881
1896	+ 4,160	+ 716	+123	+ 5,069
1897	+ 5,380	+ 556	+107	+ 6,043
1898	+ 6,884	+ 406	+ 85	+ 7,475
1899	+ 809	+ 384	+139	+ 1,332
1900	+11,513	+ 638	+168	+12,319
1901	+ 271	+ 301	+120	+ 692
1902	+ 1,574	− 107	+149	+ 1,616
1903	+ 2,934	+ 75	+114	+ 3,123
1904	+ 3,465	− 251	+ 78	+ 3,292
1905	+ 619	− 93	+100	+ 626
1906	+ 2,054	+ 233	+185	+ 2,472
1907	+ 7,745	+ 151	+228	+ 8,124
1908	+ 3,358	+ 420	+150	+ 3,928
1909	− 1,601	+ 436	+101	− 1,064
1910	−15,366	+ 401	+129	−14,836
1911	+16,092	+ 373	+120	+16,585
1912	+10,024	+1,464	+308	+11,796
1913	− 5,008	+1,385	+301	− 3,322

* Source: *Statistical Abstract for the United Kingdom.*

CHART V

ANNUAL MOVEMENTS OF SILVER AND BRONZE COIN IN CIRCULATION IN ENGLAND, 1881–1913

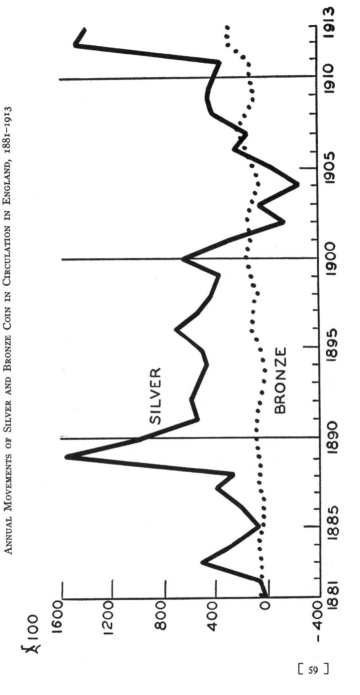

changes in specie holdings, and the resultant figure measures the amount of the internal movement.[1] This is the method which has been used for many years to determine the internal flow, particularly with respect to seasonal movements.[2] A quarterly

[1] There are six possible cases which arise in actual calculation, and these are illustrated in the table below:

	Differences between Specie Holdings of the Bank of England	Net Foreign Gold Movement for Intervening Week	Net Internal Flow for Week
1	− 100	− 200	+ 100
2	− 100	− 50	− 50
3a	− 100	+ 150	+ 250
3b	− 100	+ 50	+ 150
4a	+ 100	− 200	+ 300
4b	+ 100	− 50	+ 150
5	+ 100	+ 200	− 100
6	+ 100	+ 50	+ 50

+ = to Bank of England   − = from Bank of England

[2] In the *Economist* of January, 1882, p. 105, the device is used to show movements over a period of two years.

The following statement is typical of the general use made by the financial press: "The stock of coin and bullion is only £8,000 less than last week, although over £360,000 was taken for export, coin having, therefore, returned from the provinces." *Statist*, February 21, 1880.

See also W. Stanley Jevons, "On the Frequent Autumnal Pressure in the Money Market, and the Action of the Bank of England," read before the Statistical Society of London, on April 17th, 1866. I have used the reprint in Jevons, *Investigations in Currency and Finance*, London, 1884, pp. 160–193. One point developed is of particular interest here. He notes that the drain comes in the fall rather than in the summer when trade is most active. He attributes this to the fact that the drain comes first on country banks and private firms whose reserves are allowed to run low. They count on the payment of dividends at the quarterly period to replenish them, and it is at that time that gold is drawn from London. *Ibid.*, p. 170. Jevons notes an earlier study by William Langton in 1857. See also R. H. Inglis-Palgrave, *Bank Rate and the Money Market in England, France, Germany, Holland, and Belgium, 1844–1900*, London, 1903, esp. Ch. XIV; the *Dictionary of Political Economy*, London, 1925, pp. 73–74; and Clare, *op. cit.*, pp. 55–58. The figures of the *Dictionary of Political Economy* relating to the autumnal drain are brought up to 1912 in an appendix and are shown by the *Bankers' Magazine*, July, 1913, pp. 3–9.

One important aspect of the seasonal drain was the demand for gold on the part of the Scotch and Irish banks of issue. Under the Acts of *8th* and *9th Vict., cap. 38* (1845), these banks were allowed to continue to issue notes to the extent of their average issue during the year ending May 1, 1845. Beyond that limit notes must be issued on gold and silver coin held by them. For this purpose a return was rendered weekly, and a four-weekly average was calculated to determine if the fiduciary limit had been exceeded. When larger note issues were necessary the banks in these two countries drew gold from London to bring their average for the month up to the legal requirements. In Scotland the £1 note was used in general circulation in place of sovereigns.

CHART VI

FOUR-QUARTER MOVING AVERAGES OF NET INTERNAL CURRENCY MOVEMENTS, ENGLAND, 1881-1913

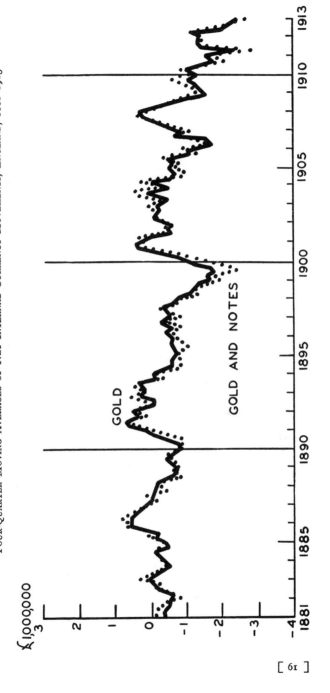

TABLE VI

QUARTERLY NET INTERIOR MOVEMENTS OF GOLD AND BANK OF ENGLAND
NOTES, ENGLAND, 1881–1913 *

(In thousands of pounds)

Quarter Ending †	Gold	Gold and Bank of England Notes	Quarter Ending	Gold and Bank of England Notes
Mar. 31, 1881 ..	+2,250	+2,735	Dec. 31........	−1,774	−1,479
June 30........	− 992	−2,111	Mar. 31, 1891 ..	+2,593	+2,833
Sept. 30........	−1,554	− 911	June 30........	− 776	−2,371
Dec. 31........	−1,285	− 487	Sept. 30........	− 418	− 180
Mar. 31, 1882 ..	+2,462	+2,807	Dec. 31........	−1,006	− 427
June 30........	−1,432	−2,337	Mar. 31, 1892 ..	+4,854	+4,966
Sept. 30........	−1,744	−1,862	June 30........	− 649	−1,806
Dec. 31........	−1,690	−1,921	Sept. 30........	−1,944	−2,035
Mar. 31, 1883 ..	+2,450	+3,766	Dec. 31........	− 220	+1,081
June 30........	− 118	− 488	Mar. 31, 1893 ..	+2,775	+2,699
Sept. 30........	−1,405	−2,542	June 30........	−1,207	−1,213
Dec. 31........	− 646	+ 348	Sept. 30........	+ 38	− 263
Mar. 31, 1884 ..	+1,615	+1,678	Dec. 31........	−1,128	−1,006
June 30........	−1,140	− 812	Mar. 31, 1894 ..	+3,213	+4,483
Sept. 30........	−2,007	−3,060	June 30........	− 510	−1,475
Dec. 31........	− 174	+1,068	Sept. 30........	−2,365	−3,285
Mar. 31, 1885 ..	+2,184	+2,275	Dec. 31........	− 846	− 402
June 30........	− 184	−1,036	Mar. 31, 1895 ..	+2,322	+2,117
Sept. 30........	−3,843	−3,246	June 30........	−1,509	−1,695
Dec. 31........	+ 205	+ 893	Sept. 30........	−2,293	−3,096
Mar. 31, 1886 ..	+3,492	+3,552	Dec. 31........	−1,027	− 382
June 30........	− 536	−1,324	Mar. 31, 1896 ..	+1,850	+1,585
Sept. 30........	− 411	− 308	June 30........	−1,178	−2,408
Dec. 31........	+ 140	+ 841	Sept. 30........	−1,981	−2,052
Mar. 31, 1887 ..	+3,367	+4,005	Dec. 31........	−1,393	− 23
June 30........	−1,112	−2,265	Mar. 31, 1897 ..	+2,939	+2,347
Sept. 30........	−1,427	− 931	June 30........	−2,287	−3,516
Dec. 31........	− 533	− 136	Sept. 30........	− 630	− 293
Mar. 31, 1888 ..	+2,303	+2,391	Dec. 31........	−2,664	−1,863
June 30........	− 436	− 804	Mar. 31, 1898 ..	+3,494	+3,239
Sept. 30........	−1,694	−3,250	June 30........	−1,180	−1,841
Dec. 31........	− 759	+ 659	Sept. 30........	−2,490	−1,741
Mar. 31, 1889 ..	+2,173	+2,158	Dec. 31........	−2,886	−2,678
June 30........	−2,180	−3,213	Mar. 31, 1899 ..	+1,686	+1,061
Sept. 30........	−2,169	−2,324	June 30........	−1,359	−1,653
Dec. 31........	− 766	+ 243	Sept. 30........	−3,044	−2,788
Mar. 31, 1890 ..	+2,729	+2,471	Dec. 31........	−3,986	−4,345
June 30........	−1,499	−2,174	Mar. 31, 1900 ..	+2,209	+1,709
Sept. 30........	−2,568	−2,371	June 30........	−2,369	−3,736

* SOURCE: Statistics of gold and note holdings of the Bank of England and net foreign gold movements compiled from the *Economist.*

† The figures for weeks which overlapped two quarters were allotted to the quarter which included four or more days.

TABLE VI (*Continued*)

Quarter Ending	Gold	Gold and Bank of England Notes	Quarter Ending	Gold	Gold and Bank of England Notes
Sept. 30........	−2,273	−2,606	June 30........	−1,958	−3,006
Dec. 31........	−2,024	−1,727	Sept. 30........	− 901	− 843
Mar. 31, 1901 ..	+3,096	+3,253	Dec. 31........	−3,588	−3,188
June 30........	+ 369	− 487	Mar. 31, 1908 ..	+4,636	+5,250
Sept. 30........	− 671	− 288	June 30........	−1,118	−2,188
Dec. 31........	−1,938	−1,649	Sept. 30........	+1,263	+1,699
Mar. 31, 1902 ..	+2,364	+3,512	Dec. 31........	−3,604	−3,797
June 30........	−1,804	−3,084	Mar. 31, 1909 ..	+3,178	+3,522
Sept. 30........	− 520	− 531	June 30........	−3,025	−3,828
Dec. 31........	−2,610	−1,903	Sept. 30........	−1,626	−1,125
Mar. 31, 1903 ..	+4,689	+5,026	Dec. 31........	−4,750	−3,902
June 30........	−1,984	−2,835	Mar. 31, 1910 ..	+3,898	+4,016
Sept. 30........	−1,211	− 442	June 30........	−2,584	−2,541
Dec. 31........	−2,062	−1,411	Sept. 30........	− 896	− 158
Mar. 31, 1904 ..	+3,637	+3,450	Dec. 31........	−5,652	−6,303
June 30........	− 565	− 459	Mar. 31, 1911 ..	+5,196	+6,003
Sept. 30........	− 448	+ 19	June 30........	−4,856	−6,474
Dec. 31........	−4,425	−4,227	Sept. 30........	−1,802	−1,435
Mar. 31, 1905 ..	+5,623	+5,843	Dec. 31........	−4,559	−4,685
June 30........	−2,930	−5,072	Mar. 31, 1912 ..	+1,382	+1,014
Sept. 30........	−1,173	− 312	June 30........	− 840	−1,255
Dec. 31........	−3,186	−3,255	Sept. 30........	−1,454	−1,066
Mar. 31, 1906 ..	+4,664	+5,512	Dec. 31........	−4,803	−4,418
June 30........	−2,297	−3,068	Mar. 31, 1913 ..	+2,938	+3,113
Sept. 30........	−4,115	−4,419	June 30........	−4,812	−5,417
Dec. 31........	−2,480	−1,851	Sept. 30........	−2,101	−2,106
Mar. 31, 1907 ..	+2,415	+2,529	Dec. 31........	−6,178	−6,151

series has been compiled from the weekly figures, and a four-quarter moving average computed.[1]

Fluctuations in the outstanding issue of the Bank of England notes have also been compiled, and Table VI includes a series combining the net quarterly movements of gold and Bank of England notes. A four-quarter moving average of this composite series is also shown in Table VII and Chart VI.[2] Another series of

[1] See Tables VI and VII and Charts II and VI. A plus sign indicates a return flow of gold or notes to the Bank of England.

[2] In addition to gold coin and Bank of England notes, country bank notes and subsidiary coins were in circulation, but the total amounts were relatively small. Variations in the country bank notes outstanding were added to the composite series of gold and Bank of England note movements, but there was little difference between this series and the series of gold and Bank of England notes alone. The fluctuations

quarterly movements of gold and Bank notes beginning with the second week of each calendar quarter was calculated. It is not presented here because the cyclical fluctuations in the two series of gold and Bank notes were similar. There were minor differences between the two series on account of the "window dressing" in connection with the preparation of company statements in January and July.

(6) It is apparent from the statistics of internal and external gold movements shown in Tables II and VI that there was a large secular increase in the amount of gold available for monetary and other uses in England. In Table VIII the net foreign movements of gold reported by the Bank of England and the London Customs House are compiled into annual series, and cumulative totals of net imports calculated. In the last column of the table the differences between the cumulative net imports are calculated. During the first five years of the period considered in this study, the Bank of England reported somewhat larger specie imports, but between 1886 and 1890 the statistics of the London Customs House show somewhat greater net imports. The differences between the two series are not large during this period. Between 1891 and 1894, however, the import of gold reported by the Customs House was much greater than that reported by the Bank, and by 1894 the cumulative net imports according to the Customs House data were nearly £24,000,000 greater than those shown by the Bank of England statistics. Again between 1897 and 1902 the net imports reported by the Customs House far exceeded the imports reported by the Bank, and the cumulative net imports of the former had grown to some £62,000,000 in excess of the Bank's net imports for the period from 1881. Aside from the year 1905, there is not a wide difference between the two sets of statistics after 1902.

The increase in the specie stock of England revealed by the cumulative imports is deceptive in part, because the foreign trade

in the country bank note issue were due principally to the downward trend in the issue. The maximum variation in the outstanding issue in any quarter was £168,000 between 1881 and 1900, and by that time the total amount had declined to about £1,250,000 from approximately £3,500,000 in 1881.

CHART VII

QUARTERLY AVERAGES OF BANK OF ENGLAND NOTES OUTSTANDING, 1881–1913

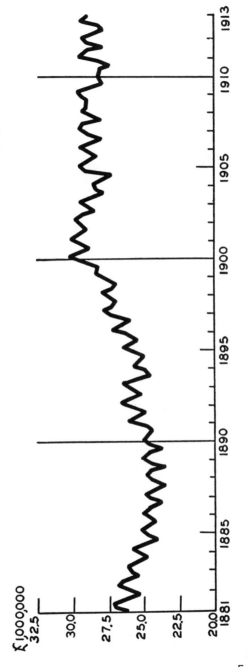

TABLE VII

FOUR-QUARTER MOVING AVERAGES OF NET INTERNAL MOVEMENTS OF
GOLD AND BANK OF ENGLAND NOTES, 1881–1913 *

(In thousands of pounds)

Centered at †	Gold	Gold and Bank of England Notes	Centered at	Gold	Gold and Bank of England Notes
July 1, 1881....	− 395	− 194	July 1.........	+ 98	− 36
Oct. 1.........	− 342	− 176	Oct. 1.........	+ 664	+ 497
Jan. 1, 1882....	− 452	− 232	Jan. 1, 1892....	+ 695	+ 638
April 1.........	− 500	− 470	April 1.........	+ 314	+ 174
July 1.........	− 601	− 838	July 1.........	+ 510	+ 552
Oct. 1.........	− 604	− 588	Oct. 1.........	− 10	− 15
Jan. 1, 1883....	− 276	− 126	Jan. 1, 1893....	− 149	+ 133
April 1.........	− 191	− 296	April 1.........	+ 346	+ 576
July 1.........	+ 74	+ 271	July 1.........	+ 120	+ 54
Oct. 1.........	− 138	− 151	Oct. 1.........	+ 229	+ 500
Jan. 1, 1884....	− 394	− 342	Jan. 1, 1894....	+ 403	+ 434
April 1.........	− 544	− 462	April 1.........	− 198	− 321
July 1.........	− 426	− 280	July 1.........	− 127	− 170
Oct. 1.........	− 284	− 132	Oct. 1.........	− 350	− 761
Jan. 1, 1885....	− 45	− 188	Jan. 1, 1895....	− 597	− 816
April 1.........	− 504	− 235	April 1.........	− 582	− 769
July 1.........	− 410	− 378	July 1.........	− 627	− 764
Oct. 1.........	− 82	+ 41	Oct. 1.........	− 745	− 897
Jan. 1, 1886....	− 170	− 31	Jan. 1, 1896....	− 662	−1,075
April 1.........	+ 688	+ 703	April 1.........	− 584	− 814
July 1.........	+ 671	+ 690	July 1.........	− 676	− 724
Oct. 1.........	+ 640	+ 804	Oct. 1.........	− 403	− 534
Jan. 1, 1887....	+ 496	+ 578	Jan. 1, 1897....	− 680	− 811
April 1.........	+ 242	+ 412	April 1.........	− 343	− 371
July 1.........	+ 74	+ 168	July 1.........	− 660	− 831
Oct. 1.........	− 192	− 235	Oct. 1.........	− 522	− 608
Jan. 1, 1888....	− 23	+ 130	Jan. 1, 1898....	− 245	− 190
April 1.........	− 90	− 450	April 1.........	− 710	− 552
July 1.........	− 146	− 251	July 1.........	− 766	− 755
Oct. 1.........	− 179	− 309	Oct. 1.........	−1,218	−1,300
Jan. 1, 1889....	− 620	− 912	Jan. 1, 1899....	−1,262	−1,253
April 1.........	− 734	− 680	April 1.........	−1,401	−1,514
July 1.........	− 736	− 784	July 1.........	−1,676	−1,931
Oct. 1.........	− 596	− 706	Oct. 1.........	−1,545	−1,769
Jan. 1, 1890....	− 424	− 446	Jan. 1, 1900....	−1,798	−2,390
April 1.........	− 526	− 458	April 1.........	−1,605	−2,244
July 1.........	− 778	− 888	July 1.........	−1,114	−1,590
Oct. 1.........	− 812	− 798	Oct. 1.........	− 792	−1,204
Jan. 1, 1891....	− 631	− 847	Jan. 1, 1901....	− 108	− 392
April 1.........	− 94	− 299	April 1.........	+ 292	+ 188

* SOURCE: Table VI.
† The calculations are made in the same manner as those in Table III.

TABLE VII (*Continued*)

Quarter Ending	Gold	Gold and Bank of England Notes	Centered at	Gold	Gold and Bank of England Notes
July 1.........	+ 314	+ 207	Oct. 1.........	− 453	− 447
Oct. 1.........	+ 31	+ 272	Jan. 1, 1908....	− 243	− 242
Jan. 1, 1902....	− 512	− 377	April 1.........	+ 298	+ 393
April 1.........	− 474	− 438	July 1.........	+ 294	+ 241
July 1.........	− 642	− 502	Oct. 1.........	− 70	− 191
Oct. 1.........	− 61	− 124	Jan. 1, 1909....	− 547	− 601
Jan. 1, 1903....	− 106	− 61	April 1.........	−1,269	−1,057
April 1.........	− 279	− 38	July 1.........	−1,556	−1,333
July 1.........	− 140	+ 84	Oct. 1.........	−1,376	−1,210
Oct. 1.........	− 402	− 310	Jan. 1, 1910....	−1,266	− 888
Jan. 1, 1904....	− 48	+ 284	April 1.........	−1,083	− 646
April 1.........	+ 143	+ 400	July 1.........	−1,308	−1,264
July 1.........	− 450	− 304	Oct. 1.........	− 984	− 750
Oct. 1.........	+ 46	+ 294	Jan. 1, 1911....	−1,552	−1,733
Jan. 1, 1905....	− 545	− 859	April 1.........	−1,778	−2,052
April 1.........	− 726	− 942	July 1.........	−1,505	−1,648
July 1.........	− 416	− 699	Oct. 1.........	−2,459	−2,895
Oct. 1.........	− 656	− 782	Jan. 1, 1912....	−1,455	−1,590
Jan. 1, 1906	− 498	− 281	April 1.........	−1,368	−1,498
April 1.........	−1,234	−1,308	July 1.........	−1,439	−1,431
July 1.........	−1,062	− 956	Oct. 1.........	−1,040	− 906
Oct. 1.........	−1,619	−1,702	Jan. 1, 1913....	−2,033	−1,947
Jan. 1, 1907....	−1,534	−1,687	April 1.........	−2,194	−2,207
April 1.........	− 731	− 793	July 1.........	−2,538	−2,640
July 1.........	−1,008	−1,127			

figures did not take account of coin carried abroad by travelers. Sovereigns were used in the British colonies and in foreign countries such as Portgual and Brazil where the sovereign was legal tender, and the net quantity of coin carried away by travelers or immigrants to these and other countries must have been considerable. In addition, the currency of Scotland and Ireland was supplied from the imports of specie shown by the statistics above. Industrial consumption accounted for a very large portion of the net imports during the period. Part of the supply for this purpose undoubtedly came from the melting down of sovereigns or of old plate, but a very large part must have come from newly imported gold.[1] The industrial consumption very likely varied somewhat

---

[1] The figure of £2,000,000 as the annual consumption of gold in the arts was accepted by the Cunliffe Committee, but this affords no light on the relative

TABLE VIII

<small>Cumulative Net Foreign Movements of Gold, 1881–1913 *</small>

(In thousands of pounds)

Net Imports +   Net Exports −

Year	Bank of England	Cumulative Difference between Imports and Exports	London Customs House	Cumulative Difference between Imports and Exports	Cumulative Differences between Net Reports of Bank of England and London Customs House †
1881	− 2,340	− 2,340	− 5,608	− 5,608	− 3,268
1882	+ 2,441	+ 101	+ 3,983	− 1,626	− 1,727
1883	+ 802	+ 903	+ 135	− 1,490	− 2,393
1884	+ 964	+ 1,867	− 350	− 1,840	− 5,707
1885	+ 1,059	+ 2,926	+ 2,037	+ 197	− 2,729
1886	− 3,980	− 1,054	− 768	− 571	+ 483
1887	+ 804	− 250	+ 715	+ 144	+ 394
1888	+ 33	− 217	+ 580	+ 724	+ 941
1889	+ 1,358	+ 1,141	+ 2,655	+ 3,379	+ 2,238
1890	+ 8,796	+ 9,937	+ 9,242	+ 12,601	+ 2,664
1891	− 1,558	+ 8,389	+ 6,063	+ 18,664	+10,275
1892	+ 1,061	+ 9,450	+ 6,668	+ 25,332	+15,881
1893	− 26	+ 9,425	+ 4,891	+ 30,223	+20,798
1894	+ 9,019	+18,444	+12,147	+ 42,370	+23,926
1895	+14,376	+32,820	+14,921	+ 57,291	+24,471
1896	− 8,099	+24,721	− 5,607	+ 51,684	+26,963
1897	− 1,040	+23,681	− 214	+ 51,470	+27,789
1898	+ 1,947	+25,628	+ 7,530	+ 59,000	+33,372
1899	+ 9,691	+35,319	+10,415	+ 69,415	+34,096
1900	+ 998	+36,317	+ 8,604	+ 78,019	+41,702
1901	+ 3,070	+39,387	+ 6,318	+ 84,337	+44,950
1902	− 246	+39,141	+ 6,311	+ 90,648	+51,507
1903	− 297	+38,844	+ 956	+ 91,604	+52,760
1904	+ 2,816	+41,660	+ 1,163	+ 92,767	+51,107
1905	+ 589	+42,249	+ 8,313	+101,080	+58,831
1906	+ 5,762	+48,011	+ 3,210	+104,290	+56,279
1907	+ 6,293	+54,304	+ 5,647	+109,937	+55,633
1908	− 3,009	+51,295	− 3,683	+106,254	+54,959
1909	+ 8,119	+59,414	+ 7,843	+114,097	+54,683
1910	+ 3,962	+63,376	+ 8,114	+122,211	+58,835
1911	+ 9,086	+72,462	+ 8,616	+130,827	+58,365
1912	+ 2,594	+75,056	+ 5,235	+136,062	+61,006
1913	+12,847	+87,903	+13,972	+150,034	+62,131

* Source: Table II.
† A minus sign indicates an excess in cumulated net imports reported by the Bank of England over cumulated net imports reported by the London Customs House.

CHART VIII

FOUR-QUARTER MOVING AVERAGES OF NET EXTERNAL GOLD MOVEMENTS AND NET INTERNAL MOVEMENTS
OF GOLD AND BANK OF ENGLAND NOTES, 1881–1913

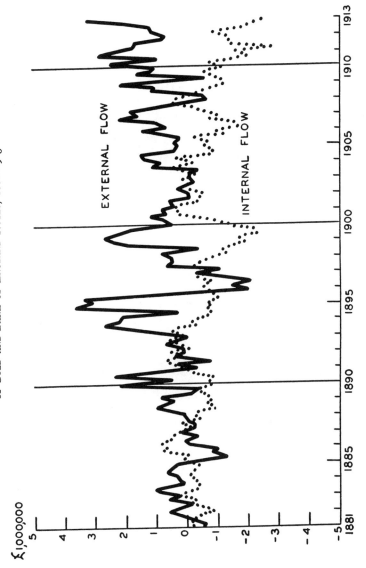

over cyclical periods, as the demand for gold ornaments is affected by cyclical conditions of business.[1] There was a considerable increase in the volume of coin held by the banks and the public,[2] and the amount held by the banks is understated somewhat in the Mint estimates.[3] The estimate of £110,000,000 for the gold supply of the United Kingdom in 1883 by Palgrave may have been somewhat high. It seems probable that there was an increase of nearly £25,000,000 in the gold circulation between 1881 and 1913.[4] When the additions to the bullion stock of the Bank of England are added, a total of nearly £40,000,000 in the gold monetary stock seems probable. In addition there must have been an increase in the Irish circulation not shown by the Mint estimates.

After allowing some £40,000,000 as the increase in the monetary stock of the country, and assuming that most of the annual industrial consumption of £2,000,000 was taken care of from newly imported specie, there is a considerable volume of gold reported by the Customs House which is not accounted for. But when further allowance is made for the stocks of bullion accumulated by the joint stock banks, an increase in the Irish circulation, and the coin and bullion lost through wear or unrecorded export, the total amount of the net import for the period seems to be accounted for without any great margin of error. The chief problem

amounts taken from imported bullion as compared with that derived from the melting down of coin or old plate. See the *First Interim Report*, p. 3. See also *Economist*, January 5, 1907, p. 12.

In 1911 the Mint estimated that only £300,000 of gold coin had been used in manufacturing during the previous 7 years. See the *Mint Report*.

This estimate is of little value for the present purpose, and it cannot be used as a basis for calculating the absorption of gold coin in the arts in earlier years.

[1] See Wesley C. Mitchell, *Business Cycles*, Berkeley, California, 1913, p. 282.

[2] See the estimates above, p. 54. The estimate of 1888 was admittedly too high, and it seems unlikely that £7,500,000 were withdrawn from the circulation between 1883 and 1888.

[3] See below, pp. 88–91.

[4] The large increase in the coin circulation in France between 1901 and 1913 suggests that this figure may be too low. Increased requirements of the public for small money units because of the increase in prices and business activity, rather than hoarding or a change in money habits, is the reason attributed by White for the expansion of coin in circulation. In this period there was a large excess of bullion entering the country which did not go into the Bank of France. The increased circulation was supplied from these imports. See White, *op. cit.*, pp. 177–182.

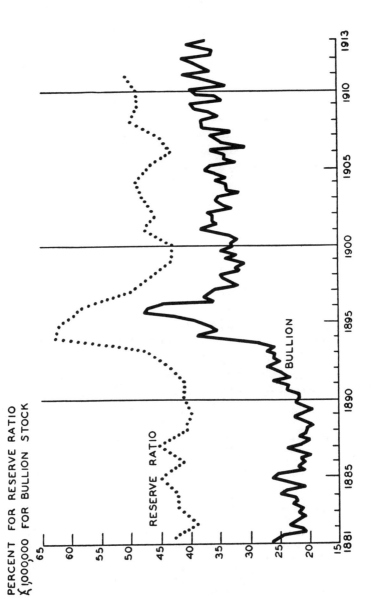

CHART IX

QUARTERLY AVERAGES OF THE BULLION STOCK OF THE BANK OF ENGLAND AND ANNUAL
AVERAGES OF THE RESERVE RATIO, 1881–1913

PERCENT FOR RESERVE RATIO
£1,000,000 FOR BULLION STOCK

RESERVE RATIO

BULLION

TABLE IX

BANK OF ENGLAND — QUARTERLY AVERAGES OF WEEKLY FIGURES,
1881–1913 *

(In thousands of pounds)

Year	Notes	Bullion	Year	Notes	Bullion
1881...........	26,100	26,400	1891...........	24,506	23,961
	26,859	26,366		24,875	23,536
	27,135	24,676		26,145	26,647
	26,237	20,876		25,510	23,160
1882...........	25,349	21,290	1892...........	24,983	24,240
	26,290	23,466		25,945	25,596
	26,872	22,459		26,646	27,270
	26,351	20,751		26,040	24,991
1883...........	25,385	22,142	1893...........	25,033	26,261
	25,847	21,130		26,179	26,236
	26,142	23,253		26,518	27,337
	25,683	22,355		25,778	25,866
1884...........	24,745	22,534	1894...........	24,583	28,640
	25,493	25,125		25,056	34,299
	25,955	23,801		25,822	39,040
	25,223	20,361		25,529	35,262
1885...........	23,995	23,393	1895...........	25,060	36,197
	24,726	26,687		25,932	37,291
	25,169	25,113		26,416	39,842
	24,621	20,827		26,091	42,473
1886...........	24,051	22,011	1896...........	25,372	47,909
	24,748	20,761		26,368	47,690
	25,147	21,370		27,414	45,663
	24,692	19,930		26,672	35,912
1887...........	23,764	22,236	1897...........	25,982	37,973
	24,620	23,852		27,502	36,531
	24,933	21,145		27,882	35,866
	24,210	20,239		27,423	31,834
1888...........	23,542	21,972	1898...........	26,930	32,882
	24,331	20,872		27,662	34,868
	24,854	20,779		27,884	35,031
	24,405	19,455		27,311	31,489
1889...........	23,471	21,536	1899...........	26,841	32,586
	24,510	22,789		27,657	30,623
	25,085	21,602		28,535	34,162
	24,461	19,712		28,479	31,705
1890...........	23,714	21,519	1900...........	28,317	35,052
	24,701	22,402		29,627	32,893
	25,054	21,541		30,154	33,368
	24,732	21,820		29,639	31,970

* SOURCE: *Statistical Abstract for the United Kingdom.*

TABLE IX (*Continued*)

Year	Notes	Bullion	Year	Notes	Bullion
1901	28,883	33,600		29,413	36,858
	29,596	35,890		29,296	32,895
	30,126	38,275	1908	28,212	37,943
	29,631	35,569		28,774	37,980
1902	29,048	36,234		29,401	37,871
	29,391	36,173		29,039	34,824
	29,870	37,591	1909	28,932	36,611
	29,316	32,798		29,396	39,077
1903	28,414	34,853		29,652	39,810
	29,033	35,662		28,878	34,159
	29,477	35,552	1910	28,136	35,547
	28,631	31,539		28,294	38,649
1904	28,021	33,995		28,382	39,915
	28,432	33,891		28,209	33,659
	28,651	36,057	1911	27,474	37,069
	28,052	33,711		28,424	38,087
1905	27,517	36,641		29,626	41,064
	29,044	37,426		28,913	36,423
	29,688	36,693	1912	28,164	38,148
	29,003	31,919		28,926	39,435
1906	28,402	34,503		29,374	41,119
	28,929	34,195		28,695	36,248
	29,558	36,684	1913	28,108	36,083
	28,677	30,389		28,591	37,562
1907	28,134	34,570		29,381	40,354
	28,889	35,350		28,879	36,284

is to account for the great difference in the net imports indicated by the two sets of statistics between 1891 and 1894 and later from 1897 to 1902. The fluctuations in the outstanding issue of Bank of England notes and the large imports of gold which did not go to the Bank of England in the first of these two periods suggest that some hoarding of currency may have taken place.[1] In the second period, the accumulation of gold by the joint stock banks may be the major explanation for the differences between the statistics of the Customs House and the Bank.

The series derived from the Mint reports are subject to a wider margin of error, since no allowance is made for coin melted down for export in the form of bullion. And the gold series shows much less regular cyclical fluctuation than the silver and bronze coin

---

[1] See below, p. 75.

series. Even the latter are subject to errors of various kinds. For example, in 1906 and 1907 a significant change in the use of bronze was brought about by the introduction of the penny slot machine on a large scale.[1] The series derived from the Mint reports also tend to lag somewhat behind those calculated on the basis of the Bank returns. One explanation for this difference is the fact that the former are based upon annual rather than quarterly statistics. But a more important reason lies in the relation of the Bank of England to the Mint. It was the practice of the Bank to meet the cash requirements of its customers — the banks primarily — from their own coin stock. Not until that ran low did they send bullion to the Mint for coinage.[2] Under these circumstances such a lag is not surprising.

(7) Bank notes do not exhibit any important cyclical fluctuation because of their limited use. This was pointed out by Mitchell in his study of business cycles,[3] and it is also indicated by the annual averages of the outstanding note issue prepared by Palgrave.[4] The quarterly averages of notes held outside the

---

[1] *Mint Report*, 1906, pp. 13–15, and 1907, p. 14.

[2] The following table taken from the *Mint Report* for 1891, p. 12, gives a good indication of this practice with respect to silver coin.

Year	Received by the Bank from the Mint £	Issued by the Bank to the Public in Excess of the Amount Received £	Received by the Bank from the Public in Excess of the Amount Issued £	Worn Coin Withdrawn from Circulation £
1877 ...	186,300	66,000	....	170,000
1878 ...	215,500	187,000	....	220,000
1879 ...	153,430	....	121,700	240,000
1880 ...	190,700	283,000	....	250,000
1881 ...	276,000	481,000	....	200,000
1882 ...	147,900	251,000	....	40,000
1883 ...	643,600	338,000	....	259,000
1884 ...	486,900	304,514	....	140,000
1885 ...	240,500	....	39,000	205,000
1886 ...	254,000	....	211,500	745,000
1887 ...	574,600	258,568	....	280,000
1888 ...	368,425	442,117	....	195,700
1889 ...	1,416,000	1,438,182	....	200,000
1890 ...	1,105,000	1,021,109	....	250,000
1891 ...	714,000	416,947	....	190,000
Total ...	6,972,855	5,487,437	372,200	2,984,700

[3] *Op. cit.*, p. 377.

[4] *Op. cit.*, p. 15.

Bank, shown in Table IX and Chart VII, reveal pronounced seasonal variation, but no regular cyclical fluctuation is apparent.[1] During the period of depression from 1891 to 1894 there was a considerable increase in the outstanding issue, although the Fiduciary Issue was increased by only £250,000 in February, 1890, and by £350,000 in January, 1894.[2] According to the Mint statistics shown in Tables IV and V, the public's requirements for silver coin declined after 1889, and the movement of bronze coin into circulation decreased after 1891. On the other hand, the circulation of gold coin was larger in 1891 than in 1890, and after a decline in 1892, there was an increase in the gold coin supplied to the public by the Mint. The fluctuations in gold coin and Bank of England notes suggest that hoarding of currency in large denominations was occurring from 1892 to 1894.

The Mint statistics of silver coin in circulation indicate a definite cyclical variation in the public demands for currency. During the years immediately preceding the crises of 1882–83, 1890, 1900, and 1907, and also in the period of expansion before 1913, there was a drain of silver from the Mint. In the periods of recession following each crisis the demands of the public were diminished. In 1902 and 1904 there was a reflux to the Mint on balance. The movement of bronze coin exhibited the same type of variation on a smaller scale. These two series are particularly significant in showing the increased need for currency during periods of high prices and business prosperity, and they reflect the conditions more clearly than the gold statistics because it was not profitable to melt down silver or bronze coin, nor is it probable that many of the coins were carried out of the country.

The gold coin series derived from the Mint Reports[3] also reflects cyclical variation in the need for currency. In 1882 the increase in circulation was slight, but in all the other periods of expansion in business there was an important increase in the demands of the public for coin. The irregularity of the series is due

[1] The increase in the issue following 1895 is due partly to the increase in the Fiduciary Issue, and partly, probably, to the rise of prices.
[2] The annual average amount of notes outstanding in 1891 was £584,000 greater than in 1890, while there was a further increase of £718,000 in 1892.
[3] See Chart IV and Table IV.

partly to the re-coinages of 1889 and 1892, since much of the coin withdrawn from circulation was not replaced.[1] The three-year moving average conforms more closely to the variations shown by the statistics of silver coin. The series of total internal movements of coin shown in Table V is dominated by the gold series, and follows the latter closely.

The two series of quarterly movements of currency derived from the Bank of England returns [2] after allowance has been made for external movements of gold are in close agreement. Since gold coin constituted the most important element in the circulation, and Bank of England notes showed no cyclical fluctuation, it is obvious that the series combining gold and notes is dominated by fluctuations in gold.

(8) In Chart VIII the four-quarter moving averages of net foreign movements of gold reported by the Bank of England and internal movements of gold and Bank of England notes are compared. The striking feature of this chart is the high inverse correlation between foreign and internal movements of gold.[3] During the period of prosperity prior to the recession of 1883–1886 there was a noticeable drain of currency from the Bank into circulation, and at the same time an influx of gold into England from abroad. The latter continued until 1884, accompanied by a slight reflux of currency from the country to the Bank in 1883. In 1885 and 1886 foreign exports of gold were heavy, and there was a concomitant reflux of coin to the Bank from circulation in 1886. The internal drain began again in 1887 and continued until 1891. There was little movement of gold on foreign account during 1887 and 1888, but from 1889 to 1891 inclusive the imports were large. In the recession following the Baring crisis a return flow of currency to the Bank set in, and this was accompanied by exports of gold. In 1894 and 1895, however, specie imports were large; but there was little movement into circulation, and the gold remained in the

---

[1] See *Mint Report*, 1890, pp. 7–9, and 1892, pp. 5–8.

[2] See Tables VI and VII and Chart VI.

[3] The chart exaggerates the correlation during the later years because the trend movements in both series widen the gap between them. Examination of the quarterly data, however, reveals the high inverse correlation.

Bank for the most part, though much of it was exported during 1896 and 1897.

In 1897 a pronounced cyclical movement of currency into circulation commenced, and it was followed by large gold imports from 1898 to 1901. The internal drain ceased during 1901, and there was some reflux to the Bank between 1902 and 1904, coupled with relatively unimportant net movements of gold on foreign account. The internal drain of gold began again in 1904 and continued until the crisis of 1907. Large gold imports accompanied this internal drain. During the depression immediately after 1907 there was a return of currency from circulation, and specie was exported from the Bank. But in 1909 the internal drain began again and continued until 1913. As in other periods of business revival large gold imports accompanied the movement of gold into circulation.

In general gold imports became important during the latter stages of the periods of business expansion, and at the same time the volume of currency in the hands of the public was expanding. In recession the flows were reversed. Thus the Bank of England acted as a channel through which gold passed from foreign countries or the mines to the British public. During the long depression from 1891 to 1896 there was an influx of gold from abroad because of the diminution of capital exports, but in other periods of depression the effect of the reduction in foreign loans was not clearly shown in specie movements.

(9) Examination of Chart IX and Table IX indicates some cyclical fluctuation in the bullion stock of the Bank of England. In the years preceding the crises of 1882–1883 and 1890 the stock declined, and the imports of gold did not remain in the Bank but went to swell the internal circulation or to provide gold for industrial purposes. The large influx of specie during 1894 and 1895 enabled the Bank to maintain a higher general level of gold, and the stock which at times before 1890 fell below £20,000,000 never averaged less than £30,000,000 after 1894. But the same cyclical variations were apparent, and in the periods of prosperity preceding the crises of 1900 and 1907 the gold stock again declined. It is

clear that the demands of the country for coin were sufficient to
lower the gold stock of the Bank and to cause the importation of
metal during the expansion periods. And in recession the reflux
from the circulation built up the Bank's stock, and even permitted
the export of gold at times without diminishing the stock appreciably. In Chart IX the annual average Reserve Ratio of the
Bank of England [1] is shown in connection with the stock of bullion. It is evident that fluctuations in the ratio were similar to the
variations in the bullion stock. So long as the Bank attempted to

TABLE X

PROPORTION OF RESERVE TO LIABILITIES OF THE BANK OF ENGLAND *

1881	43	1892	43	1902	46
1882	39	1893	47	1903	48
1883	42	1894	63	1904	49
1884	42	1895	62	1905	47
1885	45	1896	58	1906	43
1886	41	1897	50	1907	45
1887	45	1898	46	1908	50
1888	41	1899	43	1909	49
1889	40	1900	43	1910	49
1890	41	1901	48	1911	51
1891	41				

* SOURCE: A. C. Pigou, *Industrial Fluctuations*, p. 397.

keep its ratio within a narrow range, and its action was effective
upon foreign exchange rates and gold movements, this concur-
rence in the variations in the bullion stock and the reserve ratio
was to be expected over cyclical periods.

(10) It is evident that the length and magnitude of the business
cycles were important in determining the volume of the gold and
currency movements. Following the crisis of 1882 the recession
was gradual, and the same was true after 1890. Commodity
prices reached their peak in the latter part of 1889, but the crisis
did not occur until the fourth quarter of 1890. The internal drain

[1] See also Table X. The data for the table and chart are taken from A. C. Pigou,
Industrial Fluctuations, second edition, London, 1929, p. 397.

of coin did not cease, however, until the late spring of 1891. On the other hand, after 1900, and particularly after 1907, the recessions were abrupt, and the internal drain stopped more quickly. In 1900 the lag was from nine months to a year, while in 1907 it was less.

Events around 1882 do not conform altogether to this classification. In that year the peak of business reached in the late fall was sharp, but throughout 1883 there was little decline from the newly established level. The internal drain, however, ceased early in the year, although no important reflux began until 1885. Movements in this period were of somewhat less magnitude than in later periods, and it is likely that some of the important fluctuations are obscured. Moreover, the crisis itself was not pronounced, as only four years had elapsed since the severe one in 1878.

On the whole the data seem to warrant the conclusion that in cases where the period of expansion was long-continued and the succeeding recession gradual, the internal drain was abated slowly. And where the business cycle was shorter and the recession abrupt the drain ceased more quickly. The evidence afforded by the experience of the United States also confirms this conclusion.[1] It may be said, then, that the more temperate the fluctuations of the business cycle, the greater is the lag of gold and currency movements. The implications of these conclusions will be discussed below in connection with banking policy.[2]

[1] See Appendix A.
[2] See Chapter IX.

CHAPTER IV

BANK RESERVES AND CREDIT

(1) The ultimate banking reserves of the country were held in the Bank of England for the most part, and it is possible to ascertain the fluctuations in the reserve from the weekly returns of the Bank showing the Reserve of the Banking Department. This Reserve was composed principally of Bank of England notes, plus a small amount of specie, primarily silver coin. Changes in the total specie holding of the Bank closely paralleled changes in the Reserve, because of the manner in which the provisions of the Act of 1844 were carried out. Gold was received by the Banking Department, and deposit credit or notes given in return. The specie was transferred immediately to the Issue Department, and an equivalent amount of notes given in exchange to the Banking Department. Except in cases where an increase in the Fiduciary Issue was authorized and the public did not require a similar expansion in the Bank notes held by them, the specie stock and the Reserve fluctuated in a similar manner, so far as cyclical conditions are concerned. If there had been important cyclical variations in the volume of notes held by the public, the Reserve might have shown fluctuations not revealed by the specie stock, but the amount of notes outstanding was remarkably stable so far as cyclical movements are concerned. The quarterly averages of the bullion stock of the Bank shown in Table IX and Chart IX, consequently, afford an adequate indication of the movements of the ultimate banking reserve of the country.

The returns of the Bank show figures for "other deposits," which give some idea of the volume of reserves of other banks. Earlier in the century the balances of the joint stock banks were presented separately, but after 1878 this information was no longer given.[1] It seems probable that by 1914 the balances of

---

[1] The amount of the "bankers' balances" was not given in the weekly returns of the Bank, but in a special annual report published by order of Parliament. Under the privilege of a rule of Parliamentary procedure, the item was struck from the

these banks constituted the larger part of the "other deposits," because of the amalgamation of private banks with the joint stock banks. The remainder of the "other deposits" consisted of balances of private bankers, London branches of foreign or overseas banks, and other financial houses operating in the London money market.

No regular cyclical variation in the amount of the "other deposits" was found by Persons, Silberling and Berridge, after the seasonal variation and secular trend had been removed from a monthly series of these deposits.[1] An examination of the quarterly averages of *total* deposits compiled in the *Statistical Abstract for the United Kingdom* showed no cyclical fluctuation. The absence of regular variations in deposits of the Bank in connection with business fluctuations tends to substantiate the contention of English bankers that changes in these deposits were associated principally with local conditions in the London money market, and had little to do with conditions of credit and prices throughout the country.

It is possible that while the "other deposits" showed little cyclical variation, the share held by joint stock and private banks may have fluctuated somewhat in accordance with business conditions, provided the portion held by other financial houses varied in an inverse manner. Information as to the constitution of the

form of the report by one of the Directors of the Bank of England, Mr. Thomson Hankey, who was also a member of the House of Commons. The apparent reason for this action at the time was that at intervals the balances of the joint stock banks exceeded the Reserve of the Bank of England. It was felt that if this situation were generally known, fear would be engendered as to the safety of general banking deposits and bank notes. How largely this was justified is an open question, since the reports were not published until some time after the situation was a thing of the past. It was stated that thereafter the "bankers' balances" did not increase, but it seems more probable that the amalgamation of private banks with joint stock banks increased the balances of the latter during the years from 1881 to 1914. An excellent summary of the question is given in the *Bankers' Magazine*, March, 1898, pp. 349 ff. See also W. J. Atchison, "On the Ratio a Banker's Cash Reserve Should Bear to his Liability on Current and Deposit Accounts, as Exemplified by the London Clearing Joint Stock Banks; and on the Relation of the Clearing Banks to the Bank of England," *Journal of the Institute of Bankers*, vol. VI, part vi (June, 1885), pp. 307–311; Palgrave, *op. cit.*, pp. 24–27 and Ch. III; and Henry W. Macrosty, "Submerged Information," *Journal of the Royal Statistical Society*, vol. XC (New Series), part II, 1927, pp. 365–368.

[1] Persons, Silberling and Berridge, *op. cit.*, p. 183.

"other deposits" is not available, but it seems unlikely that substantial shifts over cyclical periods occurred in the share of the deposits of the Bank held by joint stock or private banks engaged in extending loans to business. The reports of two joint stock banks gave the balance held at the Bank of England separately from the cash in the till, and these figures are shown in Table XI, and are plotted in Chart X. Some increase in the balance at the Bank of England is shown before each of the major crises in the case of the Union Bank of London.[1] To the extent that cash held by the banks and balances at the Bank of England were increased by reducing holdings of short-term bills, discounting at the Bank would be increased. Average deposits of dealers who kept accounts at the Bank would rise, presumably, as they borrowed more freely from the Bank. Such movements could not have been very important, however, or there would have been a definite cyclical variation in the total of the "other deposits." Nor does it seem probable that other financial houses having deposits at the Bank would permit them to be drawn down to any great extent during periods of business prosperity, since demands for credit by their customers would increase at the same time that loans made by joint stock and private banks were growing.

The fluctuations in the balance at the Bank of England are greater than the variations in the cash held on the premises in the case of the two banks for which data are available. But there seems to have been a tendency for the fluctuations to offset one another throughout much of the period. The two parts of the reserve were roughly equal in the case of the Union Bank, while the City Bank kept considerably more than half of its reserve in the form of a balance at the Bank of England during the years for which statistics are available. In the post-war period "cash in hand" has been somewhat greater than the balance held at the Bank of England.[2] It seems likely that the amalgamation of

---

[1] Important mergers around 1900 make it impossible to determine accurately any cyclical movement at that time, and the fitting of a trend line presents great difficulty for the series.

[2] See the *Report on Finance and Industry* (*The Macmillan Report*), Cmd. 3897, London, 1931, pars. 77–79 and 367–368, and Appendix Table I. The shifting of funds from cash to balances at the Bank, or *vice versa*, may be dangerous to the Bank. See *ibid.*, par. 344, and Keynes, *op. cit.*, vol. II, p. 75.

CHART X

Cash in Hand and Balance at the Bank of England, 1881–1913

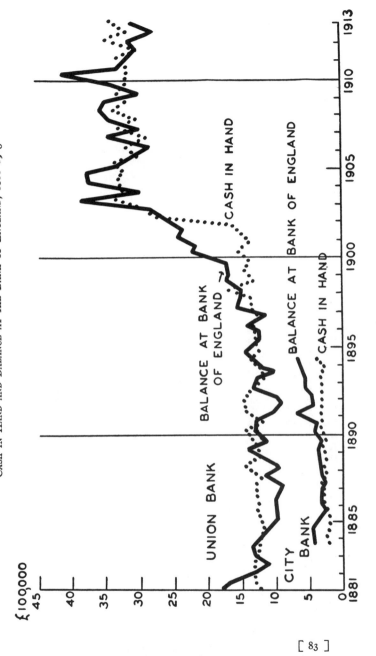

TABLE XI

Cash in Hand and at Bank of England, 1881–1913 *

(In thousands of pounds)

Date	Union Bank of London, Limited †		City Bank of London, Limited ‡	
	Cash in Hand	Cash at Bank of England	Cash in Hand	Cash at Bank of England
Dec. 31, 1881	1,213	1,813		
June 30, 1882	1,316	1,676		
Dec. 31, 1882	1,301	1,356		
June 30, 1883	1,325	1,117		
Dec. 31, 1883	1,283	1,331		
June 30, 1884	1,249	1,364	203	411
Dec. 31, 1884	1,251	1,255	247	420
June 30, 1885	1,181	1,165	234	428
Dec. 31, 1885	1,239	1,008	232	389
June 30, 1886	1,289	1,014	249	252
Dec. 31, 1886	1,262	1,014	293	337
June 30, 1887	1,266	997	286	326
Dec. 31, 1887	1,277	908	282	258
June 30, 1888	1,232	1,185	253	303
Dec. 31, 1888	1,474	959	280	324
June 30, 1889	1,243	1,242	262	355
Dec. 31, 1889	1,261	1,407	334	370
June 30, 1890	1,322	1,132	251	309
Dec. 31, 1890	1,477	1,337	354	429
June 30, 1891	1,173	1,318	293	370
Dec. 31, 1891	1,436	1,096	318	722
June 30, 1892	1,464	953	310	454
Dec. 31, 1892	1,418	1,011	318	463
June 30, 1893	1,272	1,357	306	566
Dec. 31, 1893	1,168	1,269	353	561
June 30, 1894	1,194	1,053	285	611
Dec. 31, 1894	1,347	1,317	387	653
June 30, 1895	1,210	1,471		
Dec. 31, 1895	1,278	1,242		
June 30, 1896	1,306	1,244		
Dec. 31, 1896	1,317	1,449		
June 30, 1897	1,295	1,148		
Dec. 31, 1897	1,366	1,610		
June 30, 1898	1,368	1,568		
Dec. 31, 1898	1,700	1,530		
June 30, 1899	1,363	1,746		
Dec. 31, 1899	1,440	1,678		
June 30, 1900	1,377	1,727		

* SOURCE: *Economist*, Half-yearly Banking Supplements.
† In 1902 the name was changed to the Union of London and Smiths Bank, Limited.
‡ In 1896 the City Bank was amalgamated with the London and Midland Bank, Limited.

TABLE XI (*Continued*)

Date	Union Bank of London, Limited		City Bank of London, Limited	
	Cash in Hand	Cash at Bank of England	Cash in Hand	Cash at Bank of England
Dec. 31, 1900	1,566	2,262		
June 30, 1901	1,342	2,189		
Dec. 31, 1901	1,429	2,421		
June 30, 1902	1,633	2,358		
Dec. 31, 1902	2,629	2,636		
June 30, 1903	2,873	2,808		
Dec. 31, 1903	3,361	3,866		
June 30, 1904	3,225	2,956		
Dec. 31, 1904	3,265	3,693		
June 30, 1905	3,208	3,736		
Dec. 31, 1905	3,356	3,272		
June 30, 1906	3,040	3,078		
Dec. 31, 1906	3,191	2,801		
June 30, 1907	2,868	3,398		
Dec. 31, 1907	3,298	2,955		
June 30, 1908	3,009	3,412		
Dec. 31, 1908	3,354	3,528		
June 30, 1909	3,190	3,479		
Dec. 31, 1909	3,262	2,962		
June 30, 1910	3,192	3,359		
Dec. 31, 1910	3,162	4,037		
June 30, 1911	3,151	3,275		
Dec. 31, 1911	3,312	3,171		
June 30, 1912	3,004	3,070		
Dec. 31, 1912	3,456	3,149		
June 30, 1913	3,141	2,750		
Dec. 31, 1913	3,376	3,087		

banks brought this change, rather than an alteration in banking practice, and that the two banks considered here were not typical of the banking system. Their business was largely confined to London, and till money requirements were undoubtedly less than was the case of banks at a distance. The merging of banks would increase the till cash requirements of a London bank acquiring country branches.

The reports of other banks are of little help in the determination of reserves. The joint stock banks published annual or semi-annual reports throughout the whole of the period, and in 1891 the members of the London Clearing House began to print monthly statements. Many of the private banks, however, issued no returns whatever. Fortunately the progress of the amalgamation movement brought most of them under the fold of the joint stock banks, and by 1914 the reports of the latter cover most of the banking institutions of the country, with the exception of the London branches of foreign banks. Even where available, the statements are not of the greatest value, since there was no uniformity as to the type of information presented. In many cases the item of "cash" included money lent "at call and short notice," uncleared checks, and balances at the Bank of England, as well as till money.[1] And the last included all forms of money: Bank of England notes, country bank notes, and silver and bronze coin, in addition to gold.[2] Moreover, the cash item was subject to manipulation. The banks generally increased their reserves prior to the date of publication of their returns by calling in some of their short loans from the market,[3] thereby increasing their balances at the Bank of England, and possibly also their cash held on the premises by withdrawing portions of their balances from the Bank in the form of notes or coin. This practice of "window dressing" was subjected to widespread criticism, and it is possible that it became somewhat less prevalent after 1900.[4] Some of the joint stock bank chairmen denied that their banks indulged in the practice, and maintained that they were forced to call loans only because some of their customers demanded cash in order that company statements might reveal more liquid conditions.[5] But un-

---

[1] See the yearly reviews on the progress of banking in the *Bankers' Magazine*.

[2] See H. S. Foxwell, "The Banking Reserve," a lecture delivered before the Chartered Institute of Secretaries, February 24, 1909; reprinted from the *Secretary*, March, 1909, in *Papers on Current Finance*, London, 1919, pp. 141–144, for an able discussion of the item of "cash" as reported by banks.

[3] See *Bankers' Magazine* and *Economist, passim*. The banks issued their statements as of different days of the week and this made it unnecessary to call as large a volume of loans as would have been the case if the statements had been issued on the same day. Cf. the *Macmillan Report, op. cit.*, pars. 368–370, and Keynes, *op. cit.*, vol. II, pp. 56–57.

[4] See, however, the *Macmillan Report, op. cit.*, pars. 78, 79, and 368–370. In Chart XIV below the increase in "cash" at the end of each year is very evident in the statistics of the two banks.

[5] See, e.g., the speech of the Governor of the Union Bank of London, Limited, at the half-yearly meeting in July, 1901: "The monthly balance-sheets of a bank

doubtedly the cash items of bank statements were also inflated to some extent. The various defects in the bank reports concerning the cash item makes it impossible to use them for the determination of cyclical movements in reserves.

(2) The joint stock banks increased the currency held in their vaults considerably during the period from 1881 to 1913. The increase was due in part to the amalgamation of banks, and in part to the agitation over the question of bank reserves. The growth in currency holdings was made easier by the heavy gold flow from South Africa into England after 1893, and differences in the reports of gold movements between the Bank of England and the London Customs House are explained in part by the accumulation of coin and bullion by the joint stock banks.[1]

Merging of banks brought an increase in the total reserves held. Many of the small banks kept very small reserves, and upon combination with other banks it was necessary to bring up their total to the percentage carried by the absorbing bank. It was the practice for country banks to carry a certain percentage of their deposits in cash (perhaps fifteen, though many banks kept smaller proportions). Of this 15 per cent one-third would be in the till, and two-thirds with a London agent. The latter kept only a reserve of 15 per cent against this deposit. Thus on the basis of deposits of £100,000,000 the total reserve would be only £6,500,000. When banks were merged, the London banker would still maintain a 15 per cent reserve, which necessitated £15,000,000 against the same amount of deposits. If a country bank absorbed a bank in London, it also had to keep larger re-

afforded, perhaps, a truer test of its position than the half-yearly one, for at the end of each half-year some extraordinary movements of cash always took place which it was difficult to estimate." *Bankers' Magazine*, August, 1901, p. 288. Cf. also the speeches of the Chairmen of the London and Westminster Bank, Limited, January, 1883, *ibid.*, February, 1883, p. 150; the London and Yorkshire Bank, Limited, January, 1892, *ibid.*, February, 1892, p. 294; the London and County Banking Company, Limited, *ibid.*, September, 1897, p. 382; Lloyds Bank, Limited, January, 1902, *ibid.*, March, 1902, p. 467; the National Provincial Bank of England, Limited, February, 1903, *ibid.*, March, 1903, p. 479; and the London and Westminster Bank, Limited, January, 1905, *ibid.*, February, 1905, p. 461.

[1] See below, pp. 88–91.

serves to conform to the standard of the city banks. Against this loss, it saved the fees formerly paid to the agent.[1]

The second major reason for the accumulation of gold by the joint stock banks was the agitation for larger reserves of gold in England. The controversy upon this question waxed strongly at various times during the period — chiefly during the prosperity phases of cycles. In general participants agreed that larger reserves were needed, but no one was willing to assume the expense of getting them. And the discussion waned during the periods of depression when gold reserves were large. Obviously this was the time to secure additional gold in excess of the supplies which accumulated in such periods, but no efforts were made then. Some progress was attained, however, particularly after the crisis of 1890, and again in the years immediately preceding the great war.

In the first case the joint stock banks began to hold larger "cash" reserves after the admonitions of Lord Goschen in 1891.[2] There is no question but that more "cash" was held, but it is difficult to say whether the banks merely increased their balances at the Bank of England,[3] or actually carried more specie in their

---

[1] The large increase in cash shown in the table below is explained to a considerable extent by these changes:

PRIVATE AND JOINT-STOCK BANKS IN ENGLAND AND WALES

Year	Cash (In thousands of pounds)	Percentage of Cash to Deposits *
1833 (end)	54,846	19.3
1891	107,357	23.8
1896	163,748	29.1
1901	163,631	25.8
1906	186,801	26.6
1911	218,820	27.4
1913	242,924	27.9
1914	284,969	29.9

* Erroneously printed as Capital.

These figures are extracted from a table in Joseph Sykes, *The Amalgamation Movement in English Banking, 1825–1924*, London, 1926, p. 124.

Cf. Jackson, *op. cit.*, pp. 8–9, and *Bankers' Magazine*, February, 1898, p. 290.

[2] At Leeds in February, and before the London Chamber of Commerce in December. The first of these is reprinted in G. J. Goschen, *Essays and Addresses on Economic Questions*, London, 1905, pp. 105–127.

[3] The *Economist* comments upon the increased bankers' balances held at the Bank of England, and maintained that the currently accepted figure of £12,000,000

own vaults. The statistics for the two banks shown in Table XI indicate that after 1890 the City Bank increased its balance at the Bank of England without materially altering its "cash in hand," and the same phenomenon is to be observed in the case of the Union Bank from 1899 to 1902. Increases in the balances held at the Bank threw the burden of keeping larger gold stocks upon the latter, and no increase in the Bank's earning assets could occur unless there was assurance that the "bankers' balances" would not be drawn down again shortly.

The agitation over banking reserves diminished during the nineties, only to break out afresh around 1900. And this renewal seems to have produced more results. References are found to the effect that larger reserves were being kept by the joint stock banks thereafter.[1] But not until after 1907 is it clear that specie stocks of banks were increased to any great extent.[2] The London and Midland Bank under the guidance of Sir Edward Holden was the leader in this movement, and frequent references are found in his speeches that gold *bullion*, as well as coin, was kept by the bank.[3]

for them was not the normal amount previously held. "That at present they reach that total we do not doubt, because since the collapse of the Barings the banks have very rightly thought it prudent to strengthen their position, but that as large balances as those now held are habitually kept, is improbable." February 7, 1891, p. 167.

The *Economist* on several occasions admonishes the Bank of England for not maintaining a higher Reserve, now that the banks kept larger ones at the Bank. See *Economist*, 1891, *passim*.

The Governor of the Bank evaded a question concerning the reputed increase in the bankers' balances at the meeting in the spring of 1893. See the report of the meeting in the *Bankers' Magazine*, April, 1893, p. 611.

[1] In 1900 the Chairman of Parrs' Bank, Limited, spoke of the increased reserves held by the joint stock banks, and in 1902 the *Bankers' Magazine* called attention to the generally accepted opinion that the banks kept more cash reserves. But in neither case was it explicitly stated that more gold was held. See *Bankers' Magazine*, March, 1900, p. 483, and November, 1902, p. 601.

[2] The excess of imports of gold was not accounted for by the receipts of the Bank of England in the years following 1907, and it was maintained that they had gone into the vaults of the joint stock banks to some extent. See the speech of Sir Felix Schuster at the meeting of the Union of London and Smiths Bank, Limited, in January, 1910, *ibid.*, March, 1910, p. 477.

At the meeting in January, 1913, the Chairman of Lloyds Bank, Limited, referred vaguely to an increase of gold holding by the joint stock banks. *Ibid.*, March, 1913, p. 472. See also the speech of the Chairman of Barclay and Company in July, 1911, *ibid.*, September, 1911, p. 437.

[3] At the meeting of the Midland Bank in January, 1907, Sir Edward Holden said:

In 1914 he announced that he had been authorized by the Board of Directors to publish the holdings of the bank,[1] and in December the statement showed £8,000,000 held in coin.[2]

The reports of the Royal Mint also indicate a great increase in the gold held by the joint stock banks. Beginning in 1907 these banks co-operated in giving to the Mint the amounts of coin held on a specified date.[3] They had begun the reporting of similar statistics with regard to silver coin several years earlier. The figures for gold coin show a large increase, especially after 1910. The number of banks reporting changed somewhat, however, and a part of the growth was due unquestionably to the inclusion of more banks as time passed. But without doubt more gold was

"The Bank held a considerable amount of gold in their own vaults, and it was the intention of the board gradually to increase the amount." *Ibid.*, March, 1907, p. 460. At the meeting in January, 1908, he spoke generally of the increase in gold reserves at the various joint stock banks. *Ibid.*, February, 1908, p. 277.

In 1909 it was reported by the *Bankers' Magazine* that a shipment of £300,000 in the form of gold bars had been arranged by the Midland Bank, and that it would be deposited in their own vaults, and would not be sent to the Bank of England as usual when gold was received by a bank. *Ibid.*, April, 1909, p. 561. Cf. also p. 573. A second shipment of £400,000 in American eagles was reported in 1910. *Ibid.*, May, 1900, p. 728.

[1] *Ibid.*, March, 1914, p. 475.

[2] In 1915 Sir Edward Holden was more explicit. They held £8,000,000 in gold coin, which had taken several years to accumulate. Moreover, in 1913, at the time of the strike, they had melted a portion of their bar gold into sovereigns. He continued by saying: "We now have in our vaults gold beyond the amount we publish." He was clearly referring to that part of the bar gold purchased by them, but as yet unmelted. He also estimated the holding of gold by British banks as at least £50,-000,000. See *ibid.*, March, 1915, pp. 325 and 496.

Since it was reported in June, 1914, that the banks of England held £44,500,000 in gold coin, it is apparent that some bullion was also kept by them.

[3] The returns were for the 23rd of June of each year. The statistics below include the gold held by the Bank of England.

	Gold Coin held by All Banks in England	Exclusive of Bank of England
1907 (61 banks)	£33,296,802	....
1908	50,369,167	£11,000,000
1909	49,221,074	8,400,000
1910	44,214,173	1,200,000
1911	54,009,977	14,200,000
1912 (78 banks)	60,640,681	19,100,000
1913	69,524,127	31,100,000
1914	82,794,963	44,500,000

Mint Report for 1914, p. 9.

held, and less reliance placed upon the Reserve of the Bank of England after 1907.

While there was an important increase in the gold holdings of the joint stock banks, there was no significant cyclical variation in the amounts kept on the premises. It is clear that the reserves rose generally after and not before crises. In part this seems to have been due to the return of coin from circulation during the recession phases of business cycles, and in part to the agitation for larger reserves which was most apparent in crisis years. Undoubtedly the increase in reserves made it possible for banks to expand credit during the succeeding period of prosperity; but not until after 1907 is it certain that the banks themselves held bullion or coin stores of their own beyond till money requirements, and larger bank reserves in the preceding periods of depression probably merely increased the amount of gold which it was necessary for the Bank of England to hold.

While there was some cyclical fluctuation in the Bank's stock of bullion, the fluctuations of bank credit were associated principally with variations in the reserve ratio of the Bank, since currency requirements were largely responsible for the variations in the stock. An examination of Chart IX shows that there was a considerable difference in the average ratio between periods of business expansion and contraction. The ratio declined as demands for credit and currency increased during prosperity, and rose again during depression. The gold stock of the Bank of England fluctuated somewhat with the course of the business cycle, but reserves of other banks did not vary greatly on account of cyclical movements of business, apparently. For the most part, changes in the volume of bank credit seem to have been made possible by variations in the ratio of the Bank of England, and by gold movements. Except for the international flow of gold there was little slack in the banking system, and excess reserves built up in depression periods provided little possibility for the credit or currency expansion necessary in the succeeding periods of prosperity.

(3) Data relating to the volume of bank credit outstanding are even more difficult to secure than figures for reserves. The state-

TABLE XII

CURRENT AND DEPOSIT ACCOUNTS OF JOINT STOCK BANKS IN ENGLAND AND WALES, 1881–1913 *

(In thousands of pounds)

Date		Date	
June 30, 1881	233,900	Dec. 31	527,248
Dec. 31	241,500	June 30, 1898	541,309
June 30, 1882	246,600	Dec. 31	548,552
Dec. 31	257,300	June 30, 1899	571,006
June 30, 1883	259,500	Dec. 31	566,047
Dec. 31	269,400	June 30, 1900	571,679
June 30, 1884	280,200	Dec. 31	586,726
Dec. 31	284,000	June 30, 1901	581,601
June 30, 1885	296,000	Dec. 31	584,842
Dec. 31	294,200	June 30, 1902	583,983
June 30, 1886	297,900	Dec. 31	600,333
Dec. 31	299,200	June 30, 1903	593,248
June 30, 1887	306,400	Dec. 31	588,488
Dec. 31	308,129	June 30, 1904	583,011
June 30, 1888	309,900	Dec. 31	602,512
Dec. 31	333,500	June 30, 1905	613,176
June 30, 1889	345,600	Dec. 31	627,529
Dec. 31	352,100	June 30, 1906	627,740
June 30, 1890	356,500	Dec. 31	647,889
Dec. 31	368,700	June 30, 1907	655,351
June 30, 1891	380,700	Dec. 31	648,596
Dec. 31	391,900	June 30, 1908	655,945
June 30, 1892	395,900	Dec. 31	674,660
Dec. 31	396,638	June 30, 1909	678,664
June 30, 1893	393,132	Dec. 31	685,040
Dec. 31	393,587	June 30, 1910	706,874
June 30, 1894	402,034	Dec. 31	720,687
Dec. 31	419,026	June 30, 1911	733,757
June 30, 1895	438,866	Dec. 31	748,641
Dec. 31	455,561	June 30, 1912	752,403
June 30, 1896	499,112	Dec. 31	772,878
Dec. 31	495,233	June 30, 1913	....
June 30, 1897	508,289	Dec. 31	809,350

* SOURCE: *Economist*, Half-yearly Banking Supplements.

CHART XI

DEPOSITS OF JOINT STOCK BANKS IN ENGLAND AND WALES, 1881–1913

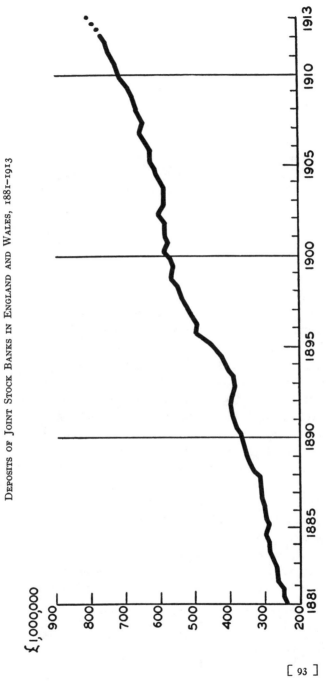

ments of all reporting banks are gathered in the semi-annual banking supplements of the *Economist*, and series of the more important items are compiled there. The monthly statements of the London clearing banks are available from 1891 on. The prime difficulty with the data is that they are not all-inclusive. Nor are the series homogeneous, since new banks were continually added to those reporting, due primarily to the progress of the amalgamation movement.

Of the various figures reported the deposit item is the most valuable on the whole. The chief defect is that it included both current and deposit accounts, i.e., checking and savings accounts. In some cases acceptance liability was also included. In the same way loans included both advances on securities and bills discounted. The total amounts of deposits and loans generally varied together, however, and deposits are considered here as a measure of the volume of bank credit. The *Economist* compiled a semi-annual series of deposits for all joint stock banks in England and Wales,[1] and this series is shown in Table XII and Chart XI. It has not been deemed worth while to remove the trend, since the growth is accounted for partly by the increase in business in the country, and partly by the addition of banks to the series, principally because the amalgamation movement brought banks which did not publish statements under the control of joint stock banks which made returns. Despite this defect, some cyclical fluctuation is apparent in deposits. They tended to grow at a more rapid rate during periods of prosperity and to fall off somewhat in depression. The measures of bank credit used by Professor Pigou[2] shown in Charts XII and XIII show the same cyclical variations, except that for the period 1896 to 1900 the 3-year moving average of the annual increase in the volume of bank credit declines.

A further examination has been made of two banks for which monthly data are available. These are the London and West-

---

[1] Even this series is not complete, although it became better as time passed. In 1883 Palgrave estimated the total bank deposits at £550,000,000, of which only £394,000,000 were accounted for by the banks which published statements. *Bankers' Magazine*, June, 1883, p. 573.

[2] Pigou, *op. cit.*, from charts facing pp. 144 and 168.

CHART XII

INDEX NUMBERS OF PRICES AND CREDITS, 1878-1914

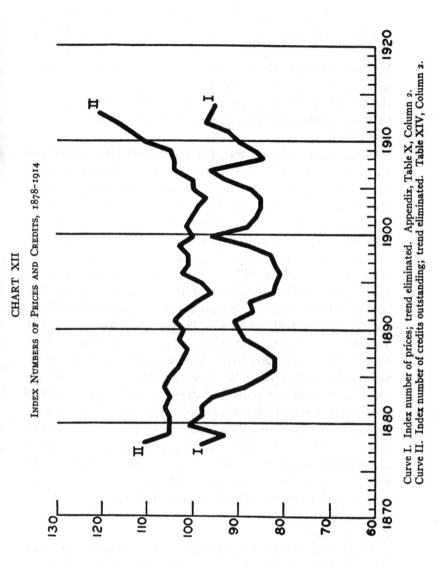

Curve I. Index number of prices; trend eliminated. Appendix, Table X, Column 2.
Curve II. Index number of credits outstanding; trend eliminated. Table XIV, Column 2.

CHART XIII

GENERAL UNEMPLOYMENT AND ANNUAL INCREASES OF
BANK CREDIT, 1880–1912

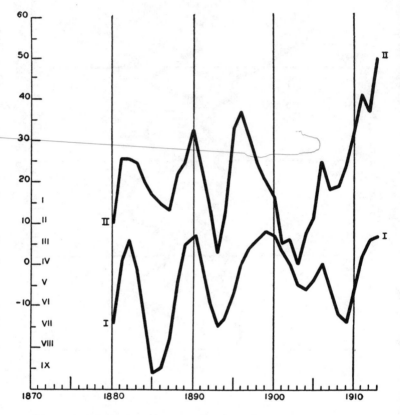

Curve I. General unemployment percentage inverted: 3-year moving
average. Appendix, Table 1, Column 6.

Curve II. Annual increases of bank credit: 3-year moving average.
Table XIII, Column 7.

[96]

CHART XIV

DEPOSIT AND CURRENT ACCOUNTS AND "CASH" RESERVES, NATIONAL PROVINCIAL BANK OF ENGLAND, LIMITED, AND LONDON
AND WESTMINSTER BANK, LIMITED, MONTHLY, 1891–1913

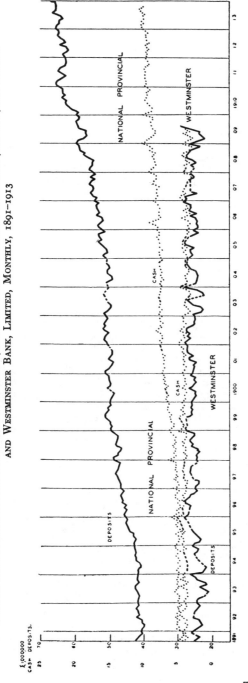

minster Bank and the National Provincial Bank. The former
dealt almost entirely with London business, while the latter had
important provincial connections in addition to its London clien-

TABLE XIII

DEPOSITS OF SELECTED ENGLISH BANKS, SEMI-ANNUALLY, 1881–1906 *

(In thousands of pounds)

Date	Provincial † banks	London ‡ Banks	Date	Provincial Banks	London Banks
Dec. 31, 1881....	24,953	37,365	June 30, 1894....	38,189	38,117
June 30, 1882....	24,908	39,011	Dec. 31........	41,199	40,166
Dec. 31........	27,801 §	38,218	June 30, 1895....	42,768	41,565
June 30, 1883....	27,352	36,109	Dec. 31........	43,164	44,987
Dec. 31........	29,118	37,218	June 30, 1896....	44,641	46,019
June 30, 1884....	28,614	40,699	Dec. 31........	45,325	40,603
Dec. 31........	28,851	33,951	June 30, 1897....	48,147	42,492
June 30, 1885....	27,677	38,290	Dec. 31........	49,124	41,389
Dec. 31........	29,925	33,478	June 30, 1898....	49,658	42,814
June 30, 1886....	29,875	34,149	Dec. 31........	50,550	43,827
Dec. 31........	29,740	34,406	June 30, 1899....	53,205	46,439
June 30, 1887....	29,995	36,425	Dec. 31........	54,622	43,764
Dec. 31........	30,678	35,023	June 30, 1900....	54,551	43,300
June 30, 1888....	30,187	36,420	Dec. 31........	55,153	43,443
Dec. 31........	33,093	35,748	June 30, 1901....	53,944	45,145
June 30, 1889....	34,183	36,951	Dec. 31........	58,116	45,412
Dec. 31........	35,418	38,071	June 30, 1902....	56,616	46,145
June 30, 1890....	35,278	36,914	Dec. 31........	60,086	43,683
Dec. 31........	35,699	37,501	June 30, 1903....	58,086	44,360
June 30, 1891....	37,848	35,743	Dec. 31........	57,415	44,764
Dec. 31........	38,318	37,611	June 30, 1904....	55,850	41,422
June 30, 1892....	38,204	36,880	Dec. 31........	59,957	43,506
Dec. 31........	38,625	35,717	June 30, 1905....	60,110	50,806
June 30, 1893....	39,451	36,319	Dec. 31........	62,655	44,292
Dec. 31........	37,611	39,700	June 30, 1906....	61,993	44,641

* SOURCE: *Economist*, Half-yearly Banking Supplements.
† This series includes six country banks: the North and South Wales Bank, Limited, the Lancashire
and Yorkshire Banking Company, Limited, the York City and County Banking Company, Limited,
the Manchester and County Bank, Limited, the Manchester and Liverpool District Banking Company, Limited, and the Birmingham District and Counties Banking Company, Limited.
‡ This series includes the London and Westminster Bank, Limited, and the London Joint Stock
Bank, Limited.
§ The York City and County Bank, Limited, was added to the series in this year. Its deposits were
£2,205,000 on December 31, 1882.

tele. While both banks grew during the period through the addition of branches, neither was concerned in important amalgamations.[1] And the item of acceptances was not large in either case.

[1] In 1909 the London and Westminster Bank was combined with the London and
County Banking Company, and the series has not been compiled after that date.

Statistics of deposits and cash reserves for the two banks are given in Table XXI and the series are plotted in Chart XIV. There is no evidence of cyclical fluctuation in the deposits of the Westminster Bank, but deposits of the National Provincial

CHART XV

DEPOSITS OF SELECTED ENGLISH BANKS, SEMI-ANNUALLY, 1881–1906

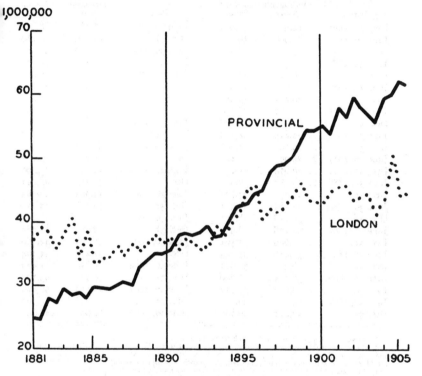

tended to increase in periods of prosperity and to remain nearly constant during depression. Cash reserves fluctuated in the same manner as the deposits, although there was a lag in the increase of reserves behind the rise of deposits of the National Provincial Bank.

A comparison has also been made of two banks whose business was principally in London — the Westminster Bank and the

London Joint Stock Bank — with a group of country banks. Since monthly statistics are not available for the latter group, semi-annual statistics of deposits have been used. They are shown in Table XIII and Chart XV. Deposits of the two London banks grew slightly between 1885 and 1889, and declined somewhat thereafter until 1893. From 1894 to 1896 they grew considerably, possibly because of the transfer of balances from country banks to London. It is not possible to determine cyclical fluctuations after 1896. Deposits of the group of country banks grew steadily throughout the period, but there was a more rapid expansion from 1887 to 1891, and again from 1894 to 1899.

Fluctuations in the volume of bank credit were very likely greater than is apparent from the crude measures used here. Shifts between current and deposit accounts were probably important over cyclical periods,[1] but accurate determination of the extent of such shifts is impossible. Variations in the rate of interest on deposits seem to indicate that time deposits increased markedly during depressions, but this evidence is not conclusive, because the deposit rate was tied in practice closely to the Bank of England rate.

Currency fluctuations appear to have been the most important forces which caused the bullion stock of the Bank of England to vary over cyclical periods, and induced gold movements on foreign account, although some variations in the volume of bank reserves occurred. Changes in the volume of bank credit undoubtedly brought some changes in the liabilities of the Bank of England, but it does not seem to have been a dominant cause of the fluctuations in the Reserve. The development of deposit banking does not seem to have brought very much slack into the monetary system, so long as the provisions of the Act of 1844 were in effect and the currency consisted largely of gold coin.

(4) Statistics of interest and discount rates are voluminous, but it is not necessary for the present purpose to make precise calcula-

---

[1] Macrosty, *op. cit.*, p. 369, estimated that two-thirds of deposits were current accounts in good times and one-half in depressions, after examination of the returns of a few banks which separated the two items.

CHART XVI

HALF-YEARLY AVERAGE MONEY RATES IN ENGLAND, 1881–1913

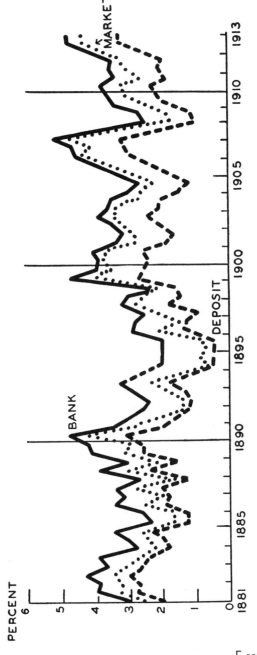

TABLE XIV

Half-Yearly Average Money Rates in England, 1881–1913 *

Year	Bank of England £	s.	d.	Market † £	s.	d.	Deposit £	s.	d.
1881	2	18	5	2	9	0	1	18	5
	4	0	4	3	7	2	2	15	8
1882	3	19	0	3	5	5	2	14	0
	4	6	3	3	9	3	3	0	6
1883	3	14	3	3	4	3	2	12	11
	3	18	10	2	17	1	2	8	10
1884	2	16	4	2	5	5	1	16	4
	3	2	4	2	10	7	1	19	3
1885	3	9	7	2	12	3	2	8	3
	2	7	1	1	13	1	1	6	4
1886	2	12	6	1	15	0	1	6	1
	3	9	7	2	10	1	1	19	7
1887	3	4	1	2	0	0	1	17	7
	3	9	7	2	16	0	2	1	5
1888	2	13	5	1	12	9	1	5	7
	3	19	6	3	3	0	2	9	6
1889	3	1	0	2	3	0	1	11	0
	4	3	0	3	4	6	2	13	0
1890	4	4	0	3	1	0	2	11	0
	4	16	0	4	5	6	3	4	6
1891	3	10	0	2	17	6	2	0	0
	3	1	0	2	7	6	1	13	4
1892	2	12	0	1	8	0	1	5	3
	2	8	6	1	9	0	1	4	0
1893	2	18	6	1	18	0	1	8	6
	3	6	0	2	7	6	1	16	0
1894		....			....			....	
	2	0	0	0	13	2	0	10	0
1895	2	0	0	0	16	9	0	10	0
	2	0	0	0	15	0	0	10	0
1896	2	0	0	0	17	6	0	10	0
	2	19	3	2	1	4	1	9	3
1897	2	16	3	1	10	0	1	5	0
	2	10	0	2	1	3	0	18	9
1898	3	5	0	2	15	6	1	15	0
	3	1	0	2	9	0	1	11	3
1899	2	2	4	2	4	9	1	12	4
	4	15	10	4	2	6	2	14	7
1900	3	19	4	3	11	2	2	11	11
	3	18	4	3	15	2	2	8	4
1901	4	1	11	3	12	3	2	11	11
	3	6	1	2	14	0	1	16	11

* Source: *Bankers' Magazine*, semi-annual statements of bankers' profit margins.
† First-class 3 months bills.

TABLE XIV (*Continued*)

Year	Bank of England £	s.	d.	Market £	s.	d.	Deposit £	s.	d
1902	3	3	4	2	15	9	1	13	4
	3	9	10	3	3	2	1	19	10
1903	3	18	3	3	8	9	2	8	3
	3	13	4	3	7	9	2	3	4
1904	3	12	0	2	14	6	2	2	0
	3	0	0	2	12	9	1	10	0
1905	2	13	9	2	5	0	1	3	9
	3	6	6	2	17	6	1	16	9
1906	3	18	7	3	10	10	2	8	7
	4	12	8	4	8	8	2	18	8
1907	4	13	10	4	3	6	3	2	10
	5	3	2	4	15	2	3	4	2
1908	3	10	0	2	17	8	1	19	6
	2	10	0	1	15	0	1	0	0
1909	2	14	3	1	17	9	1	4	3
	3	9	7	2	13	8	1	19	7
1910	3	12	7	3	2	2	2	2	7
	3	16	5	3	4	5	2	6	5
1911	3	7	9	2	14	4	1	17	9
	3	11	3	3	2	6	2	1	3
1912	3	9	3	3	5	4	1	19	3
	4	1	8	3	18	9	2	11	8
1913	4	15	11	4	6	5	3	5	11
	4	15	0	4	8	5	3	5	0

tions of cyclical movements. The main fluctuations are obvious, and the half-yearly averages compiled from the *Bankers' Magazine* are sufficient.

It is evident from an examination of the data in Table XIV and Chart XVI that rates rose during the years of prosperity, and fell off in depression. There was a tendency for the spread between the market rate and the Bank rate to narrow somewhat toward the end of the period studied. And the margin between the market rate and deposit allowances tended to increase. This was due largely to the increased concentration in banking and the cooperation of the joint stock banks with the Bank of England which resulted in keeping the deposit rate low. The market rate was in close accord with the Bank rate during periods of expansion, while in depression the spread increased.

The rates are somewhat misleading, as the Bank charged more for security loans in times of crisis, and it also got a considerable share of the fine bills on such occasions, so that market rates on the average were somewhat higher than the quoted figures for bankers' fine bills, since the latter formed a smaller proportion than usual. But the important cyclical fluctuations stand out, rates being high in periods of expansion and low in depression.

While the statistical data presented are defective in many ways, they do suggest a number of important problems which will be examined later. In the following chapters certain historical material is brought together which amplifies the previous data, and the two types of information are then considered with relation to the theory of international gold flows.

CHAPTER V

THE LONDON MONEY MARKET, 1881-1891

(1) The course of events in the money market is of especial significance in the study of foreign gold movements, for the rates of discount there are one of the primary determinants of exchange rates, which in turn induce specie flows. While too much attention is often paid to such phenomena, and too little to the fundamental conditions of credit and commodity prices, they have an important influence upon cyclical fluctuations of business. Moreover, the historical treatment which follows is particularly valuable in supplementing the preceding analysis of the statistical series, from which it was possible to draw only tentative conclusions because of the inherent defects in the data.

This survey of the money market is based largely upon material appearing in the *Economist* and the *Bankers' Magazine*. Much information is also available in the speeches of the Chairmen of the joint stock banks, however, and these reports have been utilized.[1] While contemporary opinion is not always farsighted, it is helpful in the interpretation of monetary and banking statistics, and especially in the study of trends in banking policy. In somewhat the same way, the descriptions of events in Thorp's *Business Annals* supplements the Thomas index of business conditions, and both have been used in the delineation of phases of the business cycles.

(2) The index prepared by Dr. Thomas shows that the period of revival began with the first quarter of 1881, following the depres-

---

[1] Caution must be exercised in using these speeches, for they were naturally guarded in the amount of information given. Moreover, they tended to be much alike, as the *Bankers' Magazine* facetiously remarks, ". . . there is really an opening for some enterprising person to write a typical chairman's speech, with a little space — a very little would do — for local colour, and hawk it around among directors in the same way as manuscript sermons are offered to the clergy." September, 1897, p. 336. Despite these admitted defects, some valuable information is to be found in the speeches.

sion which had existed since 1873, and accentuated by the financial distress after the failure of the City of Glasgow Bank in 1878.[1] Thorp shows the same depression existing from 1874 on, but finds that the revival began as early as the latter part of 1879.[2] He pictures 1881 and 1882 as years of "mild prosperity," and 1883 as a year of "slow recession."[3] 1884 and 1885 are characterized as years of "depression," and 1886 as "depression; slight revival."[4] There was nothing of a "boom" at any time, and the period of prosperity was short-lived.

To facilitate the description of conditions during this period, Chart XVII has been prepared. It shows the four-quarter moving averages of external gold movements reported by the Bank of England and the internal movements of gold and Bank of England notes derived from the Bank returns after allowance was made for external movements, the quarterly averages of the bullion stock of the Bank, semi-annual averages of the market and Bank rates of discount, and the Thomas index of business conditions from 1881 to the first quarter of 1891.

Cyclical movements of gold are not pronounced during the first period to be considered. From 1881 to the early part of 1883 there seems to have been more than the normal flow of coin and notes into internal circulation. This was especially marked during 1882, when the movement exceeded that of the preceding year by a considerable amount.[5] At the same time, there was a net export of gold to foreign countries throughout the earlier part of this period.[6]

In the case of foreign movements, reversal of the flow began early in 1882, under the stimulus of high Bank rates. This was counteracted to some extent by exports in the spring of the following year, when the market had broken away from the control of the Bank of England; but by May the flow had ceased, and the inward movement began again. In the case of internal movements of gold and notes, the seasonal reflux into the Bank during the chief tax-gathering period from January to March, 1883, was

[1] See Chart XVII and Table I.   [2] Thorp, *op. cit.*, pp. 167–168.
[3] *Ibid.*, p. 169.   [4] *Ibid.*, p. 170.
[5] See Table VI.   [6] See Table II.

CHART XVII

GOLD AND CURRENCY MOVEMENTS, VARIATIONS IN THE BULLION STOCK OF THE
BANK, DISCOUNT RATES, AND FLUCTUATIONS OF BUSINESS, 1881–1890

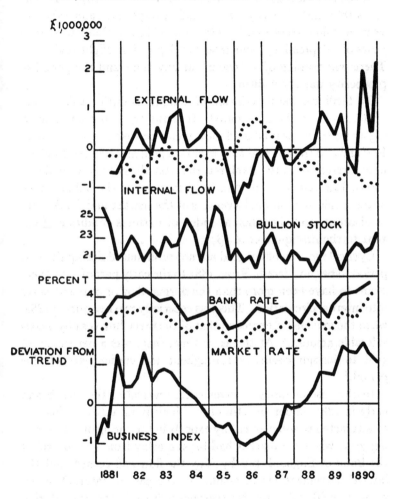

larger than in 1882. In the second quarter of the year the outflow was smaller than usual, but in the third quarter it was greater than in the preceding year. In the fourth quarter, however, there was a return flow to the Bank,[1] in contrast with the last quarters of 1881 and 1882, and it seems evident that the reversal of the outward cyclical flow may be placed near the end of the third quarter of 1883, or possibly somewhat earlier.

The fluctuations of the stock of specie at the Bank of England throw considerable light upon movements during this period. From a high point of £28,119,000 on March 23, 1881, the holdings steadily dwindled away until they were only £18,772,000 on February 1, 1882. The normal return flow from circulation in the spring combined with imports restored the stock to £24,083,000 at the end of the first quarter of 1882, but by November 1, it had again declined to £20,163,000. Despite the influx during the succeeding spring from internal circulation, the holdings attained a maximum of only £23,143,000 on February 28, 1883; and the withdrawals for foreign countries exceeded the accretion from the provinces so greatly that by May 16 the Bank held only £19,858,000, the minimum for the year.[2]

It is evident that the central cash reserve of England suffered a diminution from cyclical flows during the expansion of business from 1881 to 1883. That movements of more than usual amounts were taking place was plainly apparent to the Bank of England, and the officials made strenuous efforts to prevent undue loss. Conditions in the money market were such that control was exceedingly difficult, and it was not until the early part of 1883 that their efforts were rewarded.

Attention to the impending condition was called by the heavy

---

[1] In September the *Bankers' Magazine* observed that the 4 per cent Bank rate had been effective in increasing the reserve, especially in August, when usually money was required for local circulation. "Last year the reserve in August lost a million and a quarter; this year it has gained nearly a similar amount, and the contrast is only in part explained by the influx of gold from the continent and Australia. Four per cent. has at the same time been found effectual to check in a measure the outgoings for internal purposes, . . ." September, 1883, p. 942.

It was also pointed out that the November requirements were less than usual, and money had come in from internal circulation. *Ibid.*, December, p. 1273.

[2] These figures are taken from the official returns of the Bank. See also Chart XVII and Table IX.

withdrawals of gold for Italy and America in August, 1881.[1] The Bank raised its rate of discount to 4 and then 5 per cent, and in October it refused to sell bars.[2] The exchange finally rendered shipment of sovereigns profitable, and the drain continued. In January of 1882 there was another stringency in the money market, and the Bank rate went to 6 per cent, with a 7 per cent rate for advances. But the rate was effective for only a day or two.[3] A crisis in France caused a heavy sale of securities in London, followed by a fall of the exchange and large gold exports.[4] This movement of gold was reversed shortly, as the market rates in England advanced during the normal spring pressure, and gold flowed to the Bank of England from abroad, as well as from the internal circulation. After May, however, there was little flow either way, and the loss in the bullion stock at the Bank was due primarily to the internal drain.[5]

In general the market refused to follow the Bank during 1882.[6] At times the Bank was able to control market rates, but only at periods of seasonal pressure, and under the psychological influence

[1] Early in the year there was an unexpected stringency in the London money market due almost wholly to the large increase in stock exchange requirements and the settling day on January 14. The Bank raised its rate to 3½ per cent, the highest reached since 1879, and it charged 4 per cent for advances and 5 per cent for loans on stocks. But this stringency was short-lived, and the Bank rate fell in February to 3 per cent. On April 1st an Order in Council was obtained by the Bank to permit the increase of the fiduciary note issue by £750,000. The reserve of the Banking Department was thereby increased by that amount, and this helped to produce ease in the money market, despite the extraordinary development of the stock market. See *Economist*, Commercial History and Review of 1881, p. 5.

[2] *Bankers' Magazine*, October, 1881, p. 867.

[3] See *Economist*, Commercial History and Review of 1882, p. 5.

[4] *Bankers' Magazine*, February, 1882, p. 142. The *Bankers' Magazine* points out that the value of money acts more rapidly upon the foreign exchanges than does any adverse balance of trade. Under the 6 per cent Bank rate, gold came from New York as it flowed to France. The French market became glutted and loans were offered in Paris considerably under London rates. The result was that traders borrowed there to lend in England and the exchange rose, causing gold to flow back to England; whereupon English rates fell. "Consequently, it is the rule, that where the exchange falls to a gold-moving point, it is attributable to wide differences in the rates for money between two corresponding centres, and not to any adverse balance of trade, which acts slowly and moves gold only in exceptional cases." March, 1882, p. 168.

[5] See Tables II and VI.

[6] The spread between the average market rate and Bank rate was greater in 1882 than in 1881. See Chart XVII.

of large gold withdrawals for the Continent.[1] But in 1883 the Bank began to wield more power. In the first two months of the year there was a small inflow of gold under the influence of high Bank rates, even though market rates did not rise to the same height (1½ per cent under).[2] But in March and April under a lower rate (3 per cent) at the Bank, gold flowed abroad continuously. By this time, fortunately, the supplies of short-time capital were becoming scanty. This was true not only for the London money market, but for the provinces as well.[3] And when on April 26 the Bank announced that it would revert to its old rule of lending to bill-brokers only during the six weeks preceding the end of each quarter, and that brokers were not to rely upon being able to secure loans at other times, it speedily acquired control over the market.[4] Coin and notes began to return from circulation, and gold was induced from abroad; and the banking crisis was over.

It had been maintained during 1882 that there was no need for a high Bank rate to curb boom conditions in business, and that the only reason for it was the export of gold to foreign countries.[5] But it has been seen that the foreign loss of gold, except for minor drains, ceased early in 1882. There was a net influx from that time on to May, 1883, despite intermittent net exports for certain weeks. This accretion to the metallic stock of England did not remain in the Bank, but went out to fill the need for currency in the pockets of people or the tills of banks.[6] Thus the Bank acted

---

[1] See Table II.

[2] *Economist*, Commercial History and Review of 1883, p. 4.

[3] The *Bankers' Magazine* states that the London market was bare of funds, and ". . . that a good deal of paper is coming to London from Lancashire and other districts, where money has been in request, and bills have been remitted to London in larger quantities than usual." Money in provincial centers, therefore, was not plentiful. May, 1883, p. 548.

"Money" here clearly means liquid capital and not specie.

[4] *Economist*, Commercial History and Review of 1883, p. 4. See also *Bankers' Magazine*, June, 1883, pp. 568–569 and 614–615.

[5] "As far as the trade demand for discount has been concerned there has not all through this time been any necessity for a high rate in London. The advance [of the Bank rate] was made to arrest an outflow of gold and to check the undue absorption of securities offered here from Paris, which was running us in debt largely to that market." *Bankers' Magazine*, March, 1882, p. 263.

[6] This is clearly shown in Chart XVII.

merely as the transmitter of specie from the rest of the world to the pockets of the people of England.[1] This phenomenon is found to a greater degree in the case of later periods of expansion. The movement was comparatively small at this time, since the total cash reserve of England was very scanty in relation to the credit structure based upon it. Relatively slight changes in the requirements of other nations, which brought demands upon the gold market in London, or on the banks, and in the need for hand-to-hand currency within England, caused important changes in the gold stock, and compelled the Bank officials to take action to protect the reserves. And this they were able to do, though their control over the market was not complete by any means. Very possibly if the specie stock had been larger, the Bank's action would have occurred much later, and a greater expansion of business would have resulted before it was checked. It is pointed out later that while the Bank rate was not always effective in the London market, it did have important effects, because it was the basis upon which banks of the country made their advances; increases in overdraft rates had more effect upon industrial expansion than did changes in the money rates of the London market.[2]

(3) During 1884 and 1885 there was a fairly steady movement of coin into circulation, but nothing which could be deemed cyclical in nature. And the foreign flows of specie were not large on balance.[3] The bullion stock of the Bank was increased somewhat, and the average reserve ratio also rose.

In general the money market was not active on account of the trade depression, and its fluctuations arose largely from external flows of gold. In 1884 there was a flutter when gold was taken for Australia in February.[4] Later in the spring the panic in New York

---

[1] The *Economist* notes: "During 1882 the country has imported some 2,500,000 and more gold than it has exported, and this balance has all been sent into the Bank; yet it is remarked that no apparent increase of strength has resulted, the coin and bullion and reserve being as low as at the close of 1881." Commercial History and Review of 1882, p. 5.

[2] See Chapter VIII, below.

[3] See Table II.

[4] *Bankers' Magazine*, March, 1884, pp. 250–251.

caused a serious fall in security prices, and heavy sales to London followed.[1] The movement was so large that it raised exchange rates nearly to the English gold export point, but London discount rates rose and no important movements of specie took place. In December, however, the Bank was forced to raise its rate to 5 per cent, largely because of its failure to act sooner.[2] But for the most part rates were low and trade quiet during 1884.[3]

Rates were even lower in 1885 than in the previous year, and the banks found it very difficult to lend at times.[4] But in the third quarter of the year there was an exceptionally large outflow of gold into the internal circulation, coupled with an external drain, and the stock of specie at the Bank fell rapidly. While the outflow was arrested by the imposition of higher Bank rates, and by borrowing in the open market on the part of the Bank,[5] the latter found it very difficult to exert control because of the great plethora of liquid capital available.

In 1886, however, movements of gold were larger. Specie continued to move abroad on balance,[6] but there was a marked diminution in the drain to the country, and for the year as a whole a net influx into the Bank from that source.[7] The foreign movements were so large as to cause a net decrease in the stock at the

---

[1] "... during the crisis a number of English financial houses bought railroad stocks largely, and this, together with the action of the New York banks in sustaining one another, calmed the money market, and cash in safe directions once more became plentiful." *Ibid.*, June, 1884, p. 612.

[2] *Ibid.*, December, 1884, p. 1328. Cf. *Economist*, Commercial History and Review of 1884, p. 4. In February the Bank had relaxed the rule imposed in the previous May restricting loans to brokers to the six weeks preceding the end of each quarter. *Bankers' Magazine*, March, 1884, p. 329.

[3] The *Economist* in its review of the statements of the joint stock banks for the half-year ending on December 31, 1884, remarked on the decline of discounts and advances: "This, of course, was to be expected, not only from the known stagnation of business, but also from the hand-to-mouth policy forced upon traders by a comparatively high Bank rate, combined with an absence of domestic demand." February 21, 1885, p. 218.

[4] See, e.g., the speeches of the Chairman of the London and Westminster Bank, Limited, the London Joint Stock Bank, Limited, and the Union Bank of London, Limited, at the meetings in July, 1885. *Bankers' Magazine*, August, 1885, pp. 765, 769, and 782.

[5] See *Economist*, Commercial History and Review of 1885, pp. 4–5.

[6] See Table II.

[7] See Table VI.

Bank,[1] and the maximum holdings for the year were £22,992,000 on February 24.

The Bank exerted a certain amount of effort to prevent the large exports of gold, and at times acquired some control over the market.[2] For much of the period, however, it was ineffective, and its policy was severely condemned by the *Economist*.[3] But there is much to be said for the Bank. In the face of the reflux of gold from the interior, control over the market was difficult. The continued depression, with much unemployment, is probably the major explanation of the large flow of gold to the Bank, as coin would be drawn out of small hoards, and ultimately pass into the hands of the Bank of England. There was also a tendency for country banks to send some of their resources to London for investment at this time, and it is possible that coin and notes were drawn to London as a result.[4] The spread between the average market and Bank rate widened considerably in 1886 and the early part of 1887.

In 1886 the same passage of specie through London as a channel is found as in 1882, but the flow was in the reverse direction. Much gold also passed through London from America to the Continent.[5]

(4) According to the Thomas index, the low point of the depression following the crisis of 1883 was reached in the second quarter of 1886, and was succeeded by an almost continuous rise

[1] See *Bankers' Magazine*, June, 1886, p. 527; August, p. 642; and November, p. 951, and Chart XVII.

[2] See *Bankers' Magazine*, June, 1886, p. 527; October, pp. 875-876; and December, p. 1023. The Bank borrowed on stock in the months of September and December in order to acquire control, but it was only momentarily successful.

[3] Attention was called to the foreign drain of gold which extended throughout the year. At times the Bank ignored it, and at other times was unable to check it, due to its lack of control over the open market. It continues: "The Bank directors, in fact, seem to have been going on the principle that so long as their reserve was sufficient for internal requirements, the foreign movements might be left to regulate themselves, and, of course, the open market habitually works on the assumption that, if our stock of bullion needs looking after, it is the business of the Bank to see to it. And the result of this has been that the Bank always failed to take protective measures until the last moment, and when it found itself compelled to act, could not get the market to go with it." *Economist*, Commercial History and Review of 1886, p. 4.   [4] See below, Chapter VIII.

[5] See *Bankers' Magazine*, August, 1886, p. 642.

until the third quarter of 1890.[1] The same conditions are indicated by Thorp,[2] who marks the end of the period of expansion as November, 1890, with respect to finance, with industry and trade slackening possibly somewhat earlier.[3] The index of prices reached its peak in the last quarter of 1889.

Movements of gold were not important in 1887, but in the latter part of 1888 the internal movement of coin became greater with expanding business. It continued in greater amount during 1889 and 1890 but ceased during 1891, and a reflux began, which continued through 1894.[4] During 1889 and 1890, there was a large net influx of gold into England from abroad. The movement to the Bank of England ceased after 1891, and not until 1894 did it become important again.[5]

In 1887 the flow from the country to the Bank ceased. The Bank was somewhat better off with respect to foreign flows than in 1886, but its specie stock declined somewhat during the year. In 1888 an important internal movement began, under the impetus of business revival. The Bank was forced to take strenuous measures, especially in the fall of the year when Bank of England *notes* were taken for Russia,[6] and withdrawals of cash were made by country banks to strengthen their till reserves after the failure of a local bank.[7] By the first week in December the stock of metal at the Bank had fallen to only £18,304,000, and the Bank sold some of its government securities in an attempt to induce gold imports.[8]

In 1889 the internal drain of coin continued, but large acces-

---

[1] See Table I.   [2] *Op. cit.*, pp. 170–171.
[3] *Ibid.*, p. 171.   [4] See Chart VI and Tables VI and IX.
[5] See Chart VI and Table II.

[6] This brought another increase in the Bank rate. The *Bankers' Magazine* comments as follows: "Possibly there would have been no further rise but for the peculiar and unprecedented withdrawal of 600,000 in notes from the Bank of England, on the 13th, September, for Russia." October, 1888, p. 1094. These notes were used as a partial basis for the rouble currency, which was a fiduciary currency at the time.

[7] Some gold was taken for Argentina, and this coupled with the bank failure ". . . probably hastened the decision of the directors of the Bank, inasmuch as some cash was stated to have been drawn from London in order to fortify local banks." *Ibid.*, September, 1888, p. 976.

[8] *Ibid.*, December, p. 1286. It was said in the market that "the Bank showed its teeth."

sions of gold were received from abroad.[1] At the end of the summer, the Bank secured control over the market, largely by the sale of Treasury bills and Exchequer bonds. These issues were extraordinarily large in amount, as they were needed to pay off some of the old 3 per cent Consols which were not converted under Goschen's scheme.[2] And the Bank was able to make higher rates effective, especially as there were small shipments of gold to South America on account of loans, and the market became frightened.[3] For a time in October the release of dividends brought low market rates, and the Reserve of the Bank dropped very low;[4] but there was some improvement toward the end of the year, as assurance was given by financial houses that rather than allow an international war of rates which would result from advances in the Bank rate, they would go to some expense to bring gold in. Russia was the source for most of the gold finally secured, and it was believed that it was parted with because the Czarist government was desirous of paving the way for a conversion loan of £50,000,-000.[5] In the last month of the year the market was short of funds, probably because of borrowing on the part of the Bank.[6]

(5) There was a heavy drain of gold into circulation during most of 1890, and for the most part specie was imported, especially when the plight of the Barings became known to the Bank,

[1] See Table II and Chart XVII.   [2] *Ibid.*, September, 1889, p. 1151.
[3] *Ibid.*, October, 1889, p. 1264.   [4] *Ibid.*, November, 1889, p. 1396.
[5] *Ibid.*, November, 1889, p. 1397.
[6] *Ibid.*, December, 1889, p. 1494. See, however, *Economist* for a different version. It points out that as a result of the securing of gold from Russia, rates in the market fell, as the conclusion was jumped to that the financial houses which had worked this transaction to facilitate certain operations would see to it that the market was kept easy, and they ceased to exercise due prudence, causing the exchanges to fall until gold was exported. This finally produced the high Bank rate of 6 per cent. Commercial History and Review of 1889, p. 4. The Bank lost its control for a time because the State subvention to the local bodies was disbursed, and quickly found its way to the London market. The amount involved was about £1,800,000, and it had an appreciable effect. But gold withdrawals for South America and Portugal, and some for Germany, caused the Bank to raise its selling rate; and this proving inefficacious, it finally took the extreme step of borrowing from the market. It was a costly proceeding, but was successful in bringing market rates up. *Bankers' Magazine*, January, 1890, p. 63. See also the speech of the Chairman of the Union Bank of London, Limited, at the meeting in January, 1890. *Bankers' Magazine*, February, 1890, p. 328.

and it secured a large amount from France and Russia. In the money market conditions were chaotic. Throughout a large part of the year the Bank was able to exert a considerable amount of control over the market, as funds were becoming scanty. But there were intermittent periods when bill rates fell substantially below the Bank rate, and gold exports followed.

In January the Bank put up its rate to 6 per cent, and its charge for advances to 7 per cent.[1] The bullion stock at the Bank was very low at this time, due in part to the heavy demands for cash at Christmas, and in the early part of January for the payment of dividends.[2] The reflux of currency to the Bank during the first quarter of 1890 was larger than in 1888 and 1889, owing to the extra heavy withdrawals in the preceding year.[3]

The Bank retained control during the winter quarter,[4] despite

---

[1] *Ibid.*, February, 1890, p. 232.

[2] The precise date of the Bank return plays a considerable part in the showing which the Bank makes from week to week. In this year, for example, the return was published on the day before Christmas, and it is only to be expected that a large drop in specie holding would be shown. If the statement had been issued three or four days earlier, no such decline would have been likely. And much of the cash withdrawn would have come back to the Bank before a return was made on the twenty-eighth, for example. The same is true of the dividend-paying days in January. Cf. *Bankers' Magazine*, February, 1890, p. 232.

[3] See Table VI. Cf. *Economist*, February 1, 1890, p. 133, and February 22, p. 229. It is also pointed out, however, that this flow could only be temporary: "With trade extending, and prices and wages rising, an expansion rather than a contraction of the currency must be looked for. The 6 per cent rate must consequently be maintained until at least a million or two of gold has been drawn hither from abroad, and as yet it has shown very little attractive power." *Ibid.*, January 18, 1890, p. 65. And later it remarked that the internal flow was more marked than usual, ". . . because both trade and speculation have been checked by prolonged dearness of money." But expansion was to be expected after the close of the quarter, ". . . and the trade demand will reassert itself as soon as rates are reduced to a more moderate level. The directors, therefore, must prepare for a more than usually large outflow to the provinces later on, and it is necessary, therefore, that they should seek to strengthen their position as much as possible, while they have the opportunity." *Ibid.*, February 15, p. 197. The Scotch drain was larger and more protracted than usual according to the *Economist*, May 31, 1890, p. 685.

[4] The Governor of the Bank of England spoke of their policy at the annual meeting on March 20. He said: ". . . they had had a period of considerable difficulty owing to the very large increase in the demand for currency in this country consequent on the activity of trade, and the withdrawals for foreign countries. This had compelled the bank to use measures for the protection of its reserves which were stronger than had been used for some time, and which the directors were reluctant to employ at the particular period they were adopted, that period being associated with the receipt of taxation of a much larger degree than in other quarters of the year." He

an increase in the fiduciary issue on February 8th.[1] But in May and June the market broke away, and some gold was exported to the Continent.[2] Control was again secured in June, as the market was short of funds, and the stock exchange settlement at the end of the month was so large that it necessitated loans from the Bank at 3½ or 4 per cent.[3] This 4 per cent was an unusually high rate for July, but it went even higher at the end of the month in order to acquire gold from America. The passage of the Sherman Silver Bill brought gold imports from the United States, and the Bank took advantage of the opportunity of securing some of it.[4] Some gold was lost abroad during September, and there was a drain to the provinces, so that the stock at the Bank was reduced to only £19,137,000 on November 12, 1890. But the financial uncertainty preceding the failure of Baring Brothers had caused the Bank to take extraordinary measures to secure gold. £2,000,000 was borrowed from the Bank of France, through the good offices of the Rothschilds, and £1,500,000 was procured from Russia by the sale of bonds.[5] By the time the condition of Baring's became known, the stock at the Bank had been augmented considerably, and the action of the banks in subscribing to the guarantee fund kept the crisis from developing into a monetary panic.[6] The

pointed out that it was in the interest of the public to do so. *Bankers' Magazine*, April, 1890, p. 603.

[1] This gave the Bank £250,000 more notes. "By transferring to the Issue Department a fresh sum of £250,000 in Government securities, the Bank is able to take from that department an equivalent amount in gold, which now appears weekly in the banking department, and swells the reserve against deposit liabilities. While the reserve appears to be swollen, the total amount of coin and bullion held, which is the real reserve of the country, is still no more than moderate, and a rigid adherence to the policy of protecting the reserve, now that it is raised to a decent amount, is no less than we have a right to expect from the Bank of England." *Ibid.*, March, 1890, p. 391.

[2] *Ibid.*, June, p. 951, and *Economist*, April 12, 1890, p. 449, and May 24, p. 653.

[3] *Bankers' Magazine*, July, 1890, p. 1128.

[4] *Ibid.*, September, 1890, p. 1500.

[5] *Economist*, November 15, 1890, pp. 1437–1438.

[6] The government refused to participate in a guarantee fund for Baring's, although the Chancellor was willing to suspend the Bank Act. See the extracts from Lord Goschen's diary in Arthur D. Elliot, *The Life of George Joachim Goschen, First Viscount Goschen, 1831–1907*, London, 1911, vol. II, pp. 169–174 and 178. Goschen notes that the Rothschilds negotiated with the Bank of France to lend £3,000,000 to the Bank of England. *Ibid.*, p. 171. Only £2,000,000 were lent, however. See also

specie stock at the Bank increased, for the most part, during the succeeding weeks. In December some withdrawals were made for America, as a partial crisis occurred there,[1] but the Bank was able to avert any serious loss, and its position remained strong.

(6) During the early months of 1891, the exports of gold were large, as it was necessary to return the loan of the Bank of France. But in the second quarter the Bank secured a large amount. While it lost some specie in the last two quarters, its stock never fell below £22,000,000, with the exception of one week: a vastly better showing than in the preceding year. The large influx in the late spring was the result of an increase in the Bank rate, and a rise in the buying price for gold at the Bank.[2] The Bank was affected largely by foreign movements, especially the flow from the United States in the spring, and withdrawals for Austria-Hungary, which desired gold in order to resume specie payments.[3] Hence the Bank was forced to rebuild its bullion stock several times during the year. But it was able to exert control over the market much of the time. This was due primarily to the fact that the joint stock banks kept larger balances with it after Viscount Goschen's speech at Leeds in February.[4] The effect of this in-

the note describing the activities of Mr. Bertram Currie and the Bank in arranging for the guarantee fund, *ibid.*, pp. 283–284.

"The week has been one of intense anxiety and acute apprehension, but the alarm has really never deepened into panic, the proof being that there has been no internal drain upon the Bank of England, which is the unfailing concomitant of a true panic." The Bank was resorted to for large loans, and the "other deposits" and "other securities" showed a great increase, but there was no important withdrawal of currency for hoarding. *Economist*, November 22, 1890, p. 1465.

[1] There was heavy buying abroad as a result of the new tariff act, but the security market was disturbed owing to the British crisis, with a consequent rise of discount rates and a fall of the exchanges to the specie importing point. "The effects of the American over-commitments have no doubt been aggravated by the financial difficulties here. Masses of railroad securities held here were thrown upon the already overloaded American markets, and helped materially to cause their breakdown. . . ." Moreover, the failure of the Barings deprived a number of American banks of their usual facilities for drawing. The result of this partial breakdown was a hoarding of cash in America, and the bringing of gold from England. *Economist*, December 13, 1890, pp. 1561–1562.

[2] *Economist*, Commercial History and Review of 1891, p. 5.

[3] *Ibid.*, p. 5.

[4] The *Economist* points out that Goschen's speech and the resulting publicity had been very beneficial, ". . . and has led the joint-stock banks as a whole to increase

crease in the balances of bankers was to cut down their funds available for loans in the money market. The Bank had the use of the funds, and it could, and did, lend them in part, much to the chagrin of the banks.[1] But it could determine the rate, and thus control the activities of brokers to a greater extent than before. At one time the Bank succeeded in enlisting the support of a number of the joint stock banks in an attempt to bring up market rates, and these banks refused to lend at less than 4 per cent for a short period. It was impossible to maintain that rate in the market, and the trial was given up;[2] but it indicates a willingness upon the part of some of the banks to aid the Bank of England in a rather trying period.

Large movements of gold into circulation continued during the last quarter of 1890, and the seasonal reflux in the spring of 1891 closely approximated that of the preceding year in magnitude, though that had been unduly large.[3] But the usual outflow of gold from the Bank during the second quarter of 1891 was only half that of 1890, and the outward movement in the third quarter was very much smaller than the flow in the previous year.[4] In the case of gold and Bank of England notes much the same thing was

considerably their balance at the Bank of England." Commercial History and Review of 1891, p. 5. And see above, pp. 88–91.

"The Bank of England has not a very strong reserve of gold. Other banks are understood to keep, in the aggregate, two millions more unemployed cash at the Bank than before the recent action of public opinion, and accordingly a moderate reserve reflects as much scarcity of money in the outside market as an absolutely small reserve used to do." *Bankers' Magazine*, May, 1891, p. 824.

On the other hand, the Governor of the Union Bank of London, Limited, speaking at the meeting in January, 1892, maintained that ". . . they had not, so far, found that the proposed keeping of larger bank reserves had done much to lessen the amount of money offering in the short-loan market." *Ibid.*, February, 1892, p. 313.

[1] The Bank reserve was only slightly increased after the increase in the bankers' balances, and it was argued that: ". . . while the banks have abstained from making advances to their customers in order that they, the banks, may hold larger amounts ready to meet any sudden demand which may be made on them, and has placed these amounts with the Bank of England, the Bank of England itself has employed part at least of the increased balances of the bankers kept with it in making loans on its own account." *Ibid.*, July, 1891, p. 4.

[2] See *Economist*, May 30, 1891, p. 689, and the speech of the Chairman of the London and Westminster Bank, Limited, at the meeting in June, 1891, at which the Chairman told of the experiment. *Bankers' Magazine*, July, 1891, p. 31.

[3] See Table VI. And see *Economist*, January 10, 1891, p. 33.

[4] See Table VI.

found, though here the outward flow in the second quarter was equal to that of 1890, and did not fall off until the third quarter of the year.[1] There was a noticeable increase in the notes of the Bank outstanding between 1890 and 1891,[2] and there was some expansion in the outstanding note issue of provincial banks. Very likely the movement of country bank notes into circulation compelled the provincial banks to substitute Bank of England notes as till money.

(7) Throughout this period of expansion from 1886 to 1890 we find the same type of gold movements that were found in the previous one from 1881 to 1883. There was a large drain of specie into internal circulation, but an almost equally large influx of gold from abroad.[3] The latter commenced in 1889, very shortly after the internal drain began. And it did not cease until the crisis had been over some time. The internal flow was abated during the second quarter of 1891; but the Bank was not subjected to any real drain after the opening of the year 1891, for it received back a normal seasonal amount of coin from January to April, and withdrawals were not heavy after that time.

The Bank again acted as the agency for procuring the necessary gold from abroad to fill the pockets of the people of England.

---

[1] See Table VI.

[2] See Table IX, and above, p. 75.

[3] A letter to the Editor of the *Economist* signed F. D. pointed out that the efforts of the Bank to bring other banks into line with it in the matter of discounts was due to apprehensions that the stock of specie at the Bank, which was satisfactory at the moment, might during the autumn be reduced as it was in 1890, ". . . and by the belief that such reduction was then due to a foreign drain of gold under unduly low rates of dicount." He goes on to point out that from the end of March until the first of November, 1890, the stock fell some £5,000,000. But he shows that in this period there were gold imports of some £1,195,000 according to the Board of Trade. Thus the loss of £5,000,000 ". . . instead of being traceable to export was coincident with an actual gain to the country of gold from abroad, on balance of imports and exports, of over one million sterling."

"The inference, of course, must be, that both amounts were absorbed in the circulation of the country, owing to the improvement of trade, and evidence from the same sources would also prove that the same process is still going on."

The lesson which is drawn is that it is futile to resort to high rates to prevent what is not likely to happen under low rates, i.e., the export of gold. For such rates are a burden on business. The real remedy is to advance the price of gold when it is needed badly. *Economist*, June 13, 1891, pp. 365–366.

It was under a greater handicap than in the early eighties, for its stock of bullion was considerably smaller at the beginning of the period of expansion. Thus it was forced to take measures to protect its stock of specie even when internal industrial conditions were not "booming" by any means. And this action was responsible undoubtedly, to a large extent, for the lack of a panic in 1890.[1]

[1] In 1896 Joseph Ackland eulogized the conduct of the Directors of the Bank of England during the period preceding and during the Baring Crisis: "The years 1884–1889 witnessed a remarkable exhibition of firmness and sagacity on the part of the directors. The quietude of general business, and the scarcity of bill offerings, caused the market rate of discount to keep at low levels, while repeated demands for bullion for export threatened to deplete the reserves. The directors struggled to keep the bullion by maintaining rates sometimes 1, and even 2 per cent. above the market, while other bankers were taking away their discount business. More than once the directors borrowed money on Consols, which they did not want, in order to force the market rate up. It is a clear and worthy recognition of the duty forgotten before 1847 and 1857, and only imperfectly recognised before 1866, and met its rich reward when the Baring crisis arose in 1890: the directors were able to deal with the crisis and prevent a panic." He continues by showing how the rise of rates in 1889 and 1890 averted a panic in November, 1890. "English Financial Panics — Their Causes and Treatment," *Bankers' Magazine*, August, 1896, pp. 181 ff.

CHAPTER VI

THE LONDON MONEY MARKET DURING THE DEPRESSION OF THE NINETIES

(1) The present chapter continues the survey of conditions in the money market from the Baring Crisis to the Boer War. Movements of gold and currency, discount rates, and fluctuations in the index of business are shown in Chart XVIII. The series plotted are the same as those used for Chart XVII in the preceding chapter. Reference to the chart shows that the depression following the crisis of 1890 lasted until the second quarter of 1895, according to the Thomas index. There was a slight upturn in the index of business in the early part of 1894, but the index of prices continued to fall during that year. The revival beginning in 1895 lasted until 1900, although there was a decline in 1897 in the Thomas index.[1] Thorp dates the revival during the latter part of 1895, and classifies 1897 as prosperous, despite poor conditions in the textile industry.[2] For a part of 1900 the increased demand for war materials for the conflict in South Africa stimulated business, but rising prices and discount rates brought a decline in production and an increase in unemployment. The issue of large blocks of government securities was responsible to a considerable extent for the draining of funds from the market and the consequent high discount rates.[3] The index of business conditions reached its peak in the first quarter of 1900, a year before the close of the war.

During 1892 and 1893 internal movements of gold and notes were not large, and the Bank of England gained from the public circulation on balance. Foreign movements of gold in the quarters of the two years were offset almost precisely by counter movements in other quarters, and there was no important net flow either into or out of the country, according to the Bank of England returns. During 1894 and 1895, however, the internal movement of gold and notes from the Bank became more impor-

---

[1] See Table I and Charts I and XVIII.
[2] *Op. cit.*, p. 172.   [3] *Ibid.*, p. 173.

CHART XVIII

GOLD AND CURRENCY MOVEMENTS, VARIATIONS IN THE BULLION STOCK OF
THE BANK, DISCOUNT RATES, AND FLUCTUATIONS
OF BUSINESS, 1891–1901

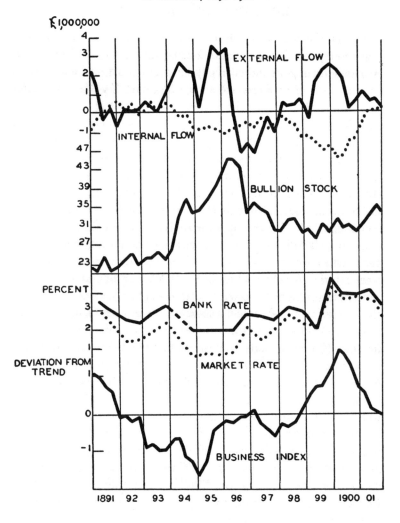

tant, and imports of specie from abroad were very large. The internal flow subsided somewhat in 1896 and 1897, and some gold was exported on balance. From 1898 to 1900 the internal movement was greater, and gold was imported from abroad.

During the years from 1891 to 1894 the imports of gold reported by the London Customs House were much larger than those reported by the Bank of England. The Bank reported net imports of £8,500,000, while the Customs House reported net imports of £21,000,000.[1] The Bank's stock of bullion increased from £23,466,000 on December 31, 1890, to £33,091,000 on January 2, 1895. It was higher than this latter figure at intervals during the period, reaching nearly £40,000,000. The estimates of the gold circulation in the hands of banks and the public indicate an increase of £2,500,000 during the three years from 1892 to 1895,[2] in spite of the reflux of gold to the Bank shown by the statistics derived from the Bank returns. The increase of £10,000,000 in the Bank stock, plus an increase of £2,500,000 in circulation, leaves some £9,000,000 unaccounted for, using the Customs House figures for imports. The figure may be even larger if there was a reflux from circulation. Industrial consumption can hardly account for this sum. There was much less gold available from imports for industrial uses in the eighties, and industrial requirements might well have been less during the depression years after 1890 than in the years immediately preceding the crisis. Hoarding of gold or unreported additions of specie to the stocks of the joint stock banks may be the explanation for the large excess of imports over increases in monetary stocks for which information is available.[3]

(2) The years following 1890 are of peculiar interest, as the depression was very prolonged. The cheapest money rates in London seemed to have no effect in starting a revival, and rates of less than $\frac{1}{2}$ of 1 per cent did not stimulate trade.[4] Of course, rates charged to most borrowers were much above these low rates, but some reduction occurred in loan rates generally.

---

[1] See above, p. 64 and Table VIII.   [2] See above, p. 54.
[3] See above, p. 75.
[4] Cf. the *Macmillan Report*, par. 234, and Keynes, *op. cit.*, vol. II, pp. 164–170.

During the spring of 1892 the Bank had little control over the market,[1] except at the end of March.[2] Despite the low rates, the movement of gold from England was not great. There was a tendency for the arriving parcels from mining countries to be taken for the Continent, but there were no large withdrawals.[3] In the second quarter there was no need for Bank loans even at the stock exchange clearings.[4] Despite the low discount rates at this period, gold poured in in considerable quantities,[5] and the stock of bullion and coin at the Bank showed an increase. Later in the fall, however, there was a rise in the Bank rate, as gold was taken for Austria, and Russia began to compete for some of it.[6] Market rates followed, and it was believed that the Bank was in position to prevent any important loss of specie.[7]

One influence operating to bring coin to the Bank after 1891 was the recall of light coin by the Mint. Under the Act of 1888 only pre-Victorian coins were called in, but now all lightweight money was subject to recoinage. The coin was sent in by banks generally during the following years, and lodged first at the Bank of England. In the case of country banks it is probable that new coin was sent very quickly to replace the shipments, but in the case of the London banks it is likely that a transfer of reserves from their tills to the Bank occurred.[8] It was maintained that part of the loss of control over the market by the Bank was due to this increase in its gold reserve.[9]

---

[1] *Bankers' Magazine*, April, 1892, p. 564.

[2] *Ibid.*, May, p. 741. It was reported that a shipment of gold was made to Russia by the Rothschilds out of their own vaults.

[3] *Ibid.*, July, p. 39.

[4] *Ibid.*, June, p. 916.

[5] Money rates were easy elsewhere as well. "At all the chief monetary centers it is the same . . . and stocks of gold are accumulating on all hands." *Economist*, June 11, 1892, p. 753.

[6] Russia was not only taking arriving parcels of gold, but also giving an occasional order to Rothschild's for gold from the Bank. *Bankers' Magazine*, December, 1892, p. 837.

[7] "The Bank of England is now believed to have control of the surplus which previously allowed the London market to beat down rates, in the absence of gold scares. . . ." *Ibid.*, December, 1892, p. 837.

[8] See *Economist*, February 27, 1892, p. 277, and *Bankers' Magazine*, May, 1892, p. 741.

[9] Some gold coin had evidently been held by the banks, since it could not be turned into the Bank except at a loss. Under the recoinage act they could not be

The year 1893 was more difficult for the Bank, for a severe crisis occurred in the United States, and the Austrian Government was attempting to build up a reserve of some £20,000,000 in order to safeguard the adoption of the gold standard.[1] In January exports of specie to France brought about a rise in market rates for money in England. These were induced by the efforts of French banks to strengthen themselves after the Panama scandals had become public; and the supplies were secured by the sale of securities abroad. The limit of note issue at the Bank of France was quickly reached as it extended its issue, and while the limit was soon raised, the momentary stoppage of emission seems to have checked the flow of gold from England.[2] During the early months of the year the export of gold from America supplied most of the Continental demands, though some gold left England for South American harvest needs.[3] In general the London money market was easy during the spring quarter,[4] though in April there was somewhat of a scarcity.

The second quarter of the year was marked by large gold imports, especially in June. But there was an important influx of currency into the banks of England, particularly those outside of London, due to the fears engendered by the Australian banking crisis which began in April. The stock of bullion and coin at the Bank fell considerably, and the Reserve even more, as notes were also drawn out of the Bank.[5] The Bank rate was advanced to 4

---

charged for this shortage, and naturally got rid of the money. See *Economist*, February 20, 1892, p. 241, and *Bankers' Magazine*, April, 1892, p. 564. In the review of the *Mint Report* for 1892 the *Bankers' Magazine* quotes an estimate of £1,200,000 as the reduction of cash in tills as a consequence of the recoinage. July, 1893, p. 3. See also A. S. Cobb, *Banks' Cash Reserves, Threadneedle Street — A Reply to "Lombard Street,"* London, 1891, p. 129.

The proportion of light coin turned in decreased sharply in 1893, as shown by the figures below taken from *Bankers' Magazine*, June, 1894, p. 866.

Year	Sovereigns	Percentage of Total Coins Weighed	Half Sovereigns	Percentage of Total Coins Weighed
1892	9,448,346	29.49	6,818,823	69.70
1893	3,990,419	11.71	2,375,835	31.33

[1] Cf. *ibid.*, January, 1893, p. 59.   [2] *Ibid.*, February, 1893, p. 240.
[3] *Ibid.*, March, 1893, p. 434.   [4] *Ibid.*, April, 1893, p. 578.
[5] See Table V. "There is nothing unusual in an efflux of cash into circulation between the middle of March and the middle of May; but, since the alarms which fol-

per cent in the middle of May, and market rates on the best bills were above this rate for a time, standing at $4\frac{1}{2}$ per cent.[1] These rates were very effective in inducing imports of specie, especially as gold was flowing from the United States as more silver was injected into the circulation.[2] A return of coin and notes from circulation set in, and the Bank Reserve and metal stock increased greatly, this being followed by a slackening of money rates.

But this decline was ephemeral, as Continental demands set in, and the breaking of the American crisis brought heavy gold exports to the United States. Some £4,000,000 was shipped from the Bank of England during the first two weeks of August.[3] "The Continent has furnished some supplementary supplies, but — as usual with a sudden drain of gold — the Bank of England has been called upon, first and most largely, to supply it." [4] With high money rates in England, there was a quick reflux, and by the end of September the stock at the Bank was largely restored.[5] Moreover, by that time the American panic was over and cash was returning to New York from the interior, instead of going to banks in the West as usual in September.[6] During the remainder of the year, the London market was quiet, and there were no important movements of specie.[7]

(3) 1894 and 1895 were marked by a most extraordinary stagnation in the money market, and by heavy gold imports. Only in the last quarter of 1894 was there a net export balance.[8] The

lowed the great Australian crisis, much more cash than usual has gone into 'circulation,' provincial and other banks having, no doubt, strengthened themselves." *Ibid.*, June, 1893, p. 922. Cf. *ibid.*, January, 1894, p. 36.

[1] *Ibid.*, June, 1893, p. 922.

[2] The Sherman Silver Purchase Act had not yet been repealed. "In May, a sort of panic lest these silver notes should corrupt the standard of values in America had caused Europe to sell securities to New York, thus adding to the debt to be settled by shipments of gold." *Ibid.*, July, 1893, p. 26.

[3] *Ibid.*, September, 1893, p. 401.

[4] *Ibid.*, September, 1893, p. 401.

[5] *Ibid.*, October, 1893, p. 564.

[6] *Ibid.*, October, 1893, p. 565.

[7] The drain of cash was large in December owing to the resumption of work in the Midlands, after the coal strike of the fall. The reflux into the Bank in the month was large, however, and the stock of specie and the Reserve became very great. *Ibid.*, February, 1894, p. 212.

[8] See Table II, and cf. *Economist*, Commercial History and Review of 1894, p. 4.

Bank rate was at 2 per cent from the last of February, 1894, until
September, 1896. But market rates were far below that rate, and
bills were discounted at times as low as ¼ of 1 per cent. It is remarkable that such low rates did not induce specie exports; but
money was almost equally cheap in foreign centers.[1] As England
was lending relatively little to other countries at the time, the
annual interest due her brought large payments, both in specie [2]
and in goods. The latter contributed to the fall of prices and
caused domestic producers much woe.[3]

During the early part of 1894 the Bank had some power over
the market, but it refrained from inflicting any restrictive rates.[4]
Money rates continued to be low and the banks complained that
money was almost unlendable much of the time.[5] For the remainder of the year the Bank was out of touch with the market
entirely, and at times the rates on the best short-time bills
dropped below the deposit rate of the banks.[6]

Not until the latter part of 1895 was there any important
change.[7] By that time a renewed activity was taking place on the
stock exchange, and the banks loaned important sums for the
settlements, and also purchased more securities themselves.[8]

---

[1] See *Bankers' Magazine*, February, 1895, p. 205.

[2] The *Bankers' Magazine* remarked that gold was flowing in from abroad, ". . .
as it always flows when England refrains from lending afresh, and the crush of unlent
capital has diminished current interest rates to what is, in practice, the lowest possible range." July, 1894, p. 35. In part the failure of Britons to lend was due to the
financial disorganization of the important borrowing countries. See *ibid.*, January,
1895, p. 66, and June, 1895, pp. 405–406.

[3] *Ibid.*, July, 1894, p. 36.

[4] *Ibid.*, March, 1894, p. 395.

[5] Cf., e.g., the speeches of the Chairmen of the London and Provincial Bank,
Limited, the London Joint Stock Bank, Limited, and the Union Bank of London,
Limited, at the semi-annual meetings in July, 1894. *Ibid.*, August, pp. 249, 254, and
269.

[6] See the speech of the Chairman of the Union Bank of London, Limited, in January, 1895. *Ibid.*, February, 1895, pp. 319–320. The banks ceased to allow depositors
interest at the old terms of 1 per cent under Bank rate, and gave only ½ of 1 per cent.

[7] A momentary rise occurred upon the floating of a large loan for America in
February, 1895, for it brought an increased number of bills into the discount market.
Some gold (£1,000,000) was taken for New York from the open market, but there
were no large withdrawals from the Bank for that purpose. *Ibid.*, March, 1895, pp.
390–391.

[8] ". . . it is, of course, an open secret that in the case of the purely London banks,
at least, the maintenance of profits is probably almost entirely due to the increased

There was an increase in the issue of loans, both domestic and foreign, for long-time purposes, and this absorbed considerable sums. And the operations involved in the payment of the Chinese indemnity locked up some £13,000,000 in the Bank of England.[1] These funds were expended during the following year, but for the moment the market lost control of the amount. Toward the end of the year discount rates in foreign countries began to advance, as a result of panics on the stock exchanges, especially in Paris and Vienna, and political scares arising from eastern Europe.[2] While the spring of 1896 witnessed another stagnant market, conditions were not as bad and did not continue so long as in the two previous years, and the money market may be said to have passed its dullest point with the summer of 1895.

The period from 1881 to 1891 was a difficult one for the Bank on account of the slender gold reserves of the country. The stagnation in business during the early nineties brought relief, and the great increase in the holdings of the Bank during 1894 and 1895 entirely changed the situation. The stock of specie at the Bank was enlarged, and it attained a maximum of some £49,000,000 on February 26, 1896. The large London joint stock banks began to hold larger balances with the Bank, and it is possible that they acquired some stores of gold of their own beyond till money requirements. The amalgamation movement in banking had concentrated banking power in fewer hands by this time, and the problem of control by the Bank of England was simplified to some extent. The Bank was in a much stronger position in 1896 to meet the demands of the revival in business than it was a decade earlier.

activity of dealings on the Stock Markets. . . ." *Ibid.*, July, 1895, p. 31. See also *ibid.*, September, 1895, p. 352.

[1] *Ibid.*, September, 1895, p. 351; October, 1895, p. 494; and December, 1895, p. 729. A loan was raised in Paris under the auspices of the Russian Government, but the Japanese Government, to whom the indemnity was due, preferred to have the funds in England, and it was necessary to effect the transfer. This was done largely by the sale of French Treasury bills in London, the proceeds being sent into the Bank of England, thus withdrawing the funds from the market. No gold was remitted apparently; though see *Economist*, Commercial History and Review of 1895, p. 4, for a different version. In August, 1895, the Bank lowered its buying rate for French gold *coin* to prevent its being shipped in.

[2] *Bankers' Magazine*, November, 1895, p. 602; December, p. 729; and January, 1896, p. 56.

(4) During 1896 and 1897 the exports of gold, particularly to the United States, were large, and the Bank store was reduced to an average of £30–35,000,000. The flow was reversed during the next few years, however, and large amounts were received by the Bank, though much was taken for industrial uses and to augment the holdings of joint stock banks. Imports at the Bank fell off after 1901, and there was little net movement during 1902 and 1903.

Internal flows from the Bank in response to trade needs were large from 1897 to 1900, but there was a marked break in the movement following the culmination of the prosperity phase of the the cycle similar to that found in former periods of decline. In this instance there was no important reflux of gold or notes to the Bank, but the outflow was completely abated. Internal movements were not important thereafter until the beginning of the drain which preceded the crisis of 1907.[1]

The heavy drain of gold to America, which has been noted above, arose largely from a remarkable change in the trade balance of the United States.[2] Not only was gold taken from the Bank of England, but for a period of eighteen months all the market arrivals were sold for foreign account.[3] In addition to the American demand, gold was desired by Japan to serve as reserve for its new gold standard. Some of the indemnity money paid to it by China — in London — was withdrawn in the form of specie, though most of the receipts were expended on ships and other capital goods in England.[4] Other demands for gold were made by Continental nations as usual.

Rates of discount in 1896 were little higher than in 1895, but by the end of 1897 the Bank was forced to take action. In October the rate was raised to 4 per cent. More important, however, was the action taken to make the rate effective in the market. Although the average rate of the Bank was no higher in 1897 than in 1896, market rates on the average were higher, and approached

---

[1] See Table VI and Chart II.
[2] See below, p. 193.
[3] See *Bankers' Magazine*, February, 1898, p. 233, and *ibid.*, 1896 and 1897 *passim*.
[4] *Ibid.*, January, 1897, p. 67; April, 1897, pp. 584–586; May, 1897, pp. 751–753; and June, 1897, p. 922.

more closely to the Bank rate.[1] This result was achieved largely by the increased discounting of bills by the Bank. There is some evidence to show that the Bank adopted a more energetic policy in this respect, not only at the head office, but at its branches as well.[2] It adopted two important changes in its policy in this regard. In September, 1897, it announced that it would now discount bills having a usance of 90 days.[3] The previous limit had been 60 days. And in December the Bank relaxed its old rule of not lending for periods of less than a week, and granted three-day loans.[4] The discounting of bills for relatively long periods brought forced ease to the market at the time of discount, but the maturing of the bills during the late fall of 1897 and the spring of 1898 helped the Bank in its efforts to control the market. In addition to its discounting policy, the Bank resorted to the older methods of borrowing for a month on Consols, and on other Government securities.[5]

Throughout the years from 1898 to 1900 the Bank continued to enforce its will upon the market as best it could. Its rate was advanced from time to time, and despite intermittent periods of

[1] *Ibid.*, January, 1898, p. 51.

[2] *Ibid.*, August, 1897, pp. 115–116; September, 1897, p. 363; and October, 1897, p. 466. It was maintained that the Bank lent at its branches at rates below the official Bank rate. This appears to have been the occasion referred to by J. Herbert Tritton as follows: "To make their rate effective, and stop the drain, the directors had to resort to borrowing on stock in the open market, and in doing so they were to all intents and purposes borrowing back at high rates their own money lent out at low rates — credit which they had created." *Op. cit.*, p. 103.

[3] *Bankers' Magazine*, October, 1897, p. 465.

[4] *Ibid.*, January, 1898, p. 64.

[5] *Ibid.*, November, 1897, p. 602, and December, 1897, p. 73. The action of the Bank in making its rate effective immediately upon its rise was praised by the Chairman of the London Joint Stock Bank, Limited, at the annual meeting in January, 1898. He claimed that it saved trade from the disturbance of uncertainty and more violent changes in rates. (Incidentally, he remarks that it was also a help to banking profits!) *Ibid.*, February, 1898, pp. 289–290.

One other factor which was asserted to have been of importance in stopping the drain of gold to the United States was the sale of American securities by Englishmen. "A large part of our indebtedness to that country [America] was balanced by the large sales of securities, amounting to many millions. Such movements in securities constituted a very important part in the play of international exchanges; their extent could only be estimated, and they defied all statistics, but they must not be overlooked." See the speech of the Chairman of the Union Bank of London, Limited, *Bankers' Magazine*, February, 1898, p. 311. See also *ibid.*, February, 1897, p. 215, and July, 1897, p. 27. And see below, pp. 135 ff.

lower rates, the average rate at the Bank gradually increased from 1896 on.[1] The open-market operations were continued, and during a considerable portion of the period of expansion command over the market was exerted by the Bank.[2] It also made certain important changes in its policy with regard to gold purchases. The first of these was instituted early in 1898, when funds were advanced against the deposit of collateral security free of interest to firms with the understanding that they would be repaid in *gold* in a few days. Specie was brought from Germany as a result.[3] Later in the year the Bank began to offer slightly more than the customary price of 77s. 9d. per ounce at certain times when it desired gold badly.[4] The operations were conducted through the agency of a bullion dealer[5] and much antagonism on the part of other firms was aroused.[6] This change was very significant, for it indicates a more aggressive policy on the part of the Bank in determining the movement of gold.[7] In this it was merely following the lead of Continental central banks which had for some time more or less controlled the free passage of gold to suit their convenience. The Bank of England did not go far along this course, but its new action was along the general lines employed elsewhere.

The outbreak of the Boer War in October, 1899, brought the Government into the money market for loans very shortly thereafter, and the course of the market for the two following years was

---

[1] See Table XIV and *ibid.*, January, 1899, p. 40, and January, 1900, p. 56.

[2] *Ibid.*, May, 1898, p. 726; July, 1898, p. 1; November, 1898, p. 595; January, 1900, p. 57; February, 1900, p. 237; and October, 1900, p. 478. See also a chart showing these rates, *ibid.*, February, 1900, pp. 240–241. At periods the market rate was above the Bank rate. Cf. also *Economist*, Commercial History and Review of 1898, p. 4.

[3] *Bankers' Magazine*, May, 1898, p. 727.

[4] *Ibid.*, December, 1898, p. 726.

[5] "At this price [77s. 9¼d. to 77s. 9½d.] the Bank of England has been able to secure a considerable amount, and a not uninteresting feature of the market has been the Bank's action in bidding over 77s. 9d. per oz. for the metal, not directly to the seller, but through the agency of a bullion-broker." *Ibid.*, March, 1899, p. 388.

[6] An article in the *Standard* is quoted to the effect that unless Continental demands were overfilled the other bullion dealers refused to sell to the Bank's agent. *Ibid.*, July, 1899, pp. 4–5.

[7] The Bank continued to vary its buying and selling rate for foreign and domestic coins. In 1899 it also made a special arrangement with some banks in the United States for the purchase of American eagles. *Ibid.*, August, 1899, p. 154.

guided largely by the action of the Government.[1] The Bank wielded more power through its control over the large Exchequer balances, and its control was aided by the natural fear and hesitancy on the part of bill brokers and other lenders during war times.

(5) The internal drain of coin was large during 1898 and 1899, and the efflux from the Bank continued well into 1900. References to the increased use of cash in the provinces on account of the expansion of trade are to be found at this time.[2] And banks operating in both the country and the city of London found that the demands for advances in the interior required more of their funds, and they were forced to withdraw balances from Lombard Street as a result.[3] As a consequence, the heavy imports of gold in these years made little impression on the Bank stock of gold.[4]

[1] *Ibid., passim,* esp. January, 1901, pp. 83–84.

[2] *Ibid.,* August, 1899, p. 153. See also the speech of the Chairman of the Union Bank of London, Limited, at the annual meeting in January, 1900, *ibid.,* February, 1900, p. 312. There was a reflux in the winter of 1900 as usual, but the drain began again in the spring and summer. In May a run occurred on a bank in Yorkshire, and cash was sent from London. *Ibid.,* June, 1900, p. 902.

[3] *Ibid.,* March, 1898, p. 392. See also the speeches of the Chairmen of the London and Yorkshire Bank, Limited, and the North and South Wales Bank, Limited, at the annual meetings in January, 1898, *ibid.,* March, 1898, pp. 462 and 466 respectively, and the speech of the Chairman of the Union Bank of London, Limited, in July, 1899, *ibid.,* August, 1899, p. 272.

[4] See Table IX. The utilization of gold imports to fill the void in the internal circulation caused by expanding business was noticed by the Chairmen of several of the large banks. See, e.g., the speeches of the Chairmen of the London and County Banking Company, Limited, and the Union Bank of London, Limited, at the meetings in January, 1900, *ibid.,* March, 1900, p. 471, and February, 1900, p. 313 respectively. See also *Economist,* July 28, 1900, p. 1061, where it is noted that larger reserves are being held by the London banks. The following table is presented by them in proof:

LONDON JOINT-STOCK BANKS

Liabilities (Excluding Acceptances)	Cash in Hand and at Bank	
June 30, 1899 .... £320,427,546	£44,044,848	13.7 %
June 30, 1900 320,783,165	50,841,147	15.8

Cf. also the speeches of the Chairmen of the National Provincial Bank, Limited, *Bankers' Magazine,* March, 1901, p. 479, the London and County Banking Company, Limited, and the London and Southwestern Bank, Limited, *ibid.,* March, 1900, pp. 472 and 473.

It is possible that some of the imported gold went into the vaults of the joint stock banks, but it is more probable that the increase in their "cash" was in the shape of balances at the Bank of England. Evidence afforded by the statements of

Not until 1901, when the internal drain had ceased, did the holdings of the Bank increase, except for the usual seasonal fluctuations.[1]

It was indicated above that the height of the prosperity phase of the cycle was reached during the first quarter of 1900. The internal drain of coin continued, however, throughout the year. Not until the last quarter of the year was there a change. While the outflow did not cease at this period, there was a marked diminution as compared to 1899, and the influx in the spring of 1901 was nearly twice as great as that of the preceding year.[2] It is evident that the change in the direction of the cyclical flow came some nine months after the expansion of business had reached its peak. The return flow to the Bank from circulation was not marked at any time, however, and from the standpoint of central reserves it is important to notice that the normal secular increase in internal currency needs, coupled with the slow decline in business transactions, prevented any augmentation of the Reserve of the Bank of England from the circulation of the country.[3]

the Union Bank of London, Limited, shown above indicates that the variation over short periods in the item of "cash" came in the Bank balances rather than in till money. It was probably true that the banks were holding more gold at this time, but it is likely that it was only a secular tendency, and that it was tied up closely with the amalgamation movement rather than cyclical movements of business.

[1] See Tables II and IX, and Chart XVIII.

[2] See Table VI.

[3] An influx of currency from the provinces to London was indicated by the Chairmen of some of the London banks in 1901. See the speeches of the Chairmen of Lloyds Bank, Limited, *ibid.*, March, 1901, p. 468, the London and County Banking Company, Limited, *ibid.*, p. 471, and the Union Bank of London, Limited, *ibid.*, August, 1901, p. 287. The Chairman of the London and Westminster Bank, Limited, maintained that country clients of banks had ceased to borrow so much and were maintaining credit balances, *ibid.*, March, 1901, p. 474. The *Bankers' Magazine* made the following statement: "Moreover, there have been indications throughout the month that trade is slackening in the provinces, and cash returning to London for employment in bills and securities." September, 1901, p. 372.

On the other hand, the Chairman of Lloyds Bank, Limited, at the meeting in January, 1902, maintained that the volume of trade had not decreased. The criterion established was the amount of coin required for weekly wages in their districts, and this had not decreased. Also the turnover of accounts was nearly as large as in previous years. *Ibid.*, March, 1902, p. 465. That a relatively small amount of funds was transferred to London from the provinces even through 1902 is indicated by the tightness in that market. For most of the period from 1900 through 1902 the Bank was depended upon. This aspect will be discussed more fully subsequently.

(6) One other aspect of the international situation in this period is of interest. There was a unanimity of opinion at that time that a large and important transfer of securities from Europe to the United States was taking place. The American trade balance was very favorable at this period, and it was believed that instead of taking gold in payment of the balance, securities sold in Europe previously were being repurchased in the United States.[1] In the summer of 1900 part of a British Government loan was floated in New York — a remarkable occurrence from the standpoint of British finance.[2] And during the fall of 1900 large amounts were temporarily loaned to England by the United States and Continental countries for a short time by the purchase of bills of exchange, or by holding bills until maturity instead of discounting them in London as usual.[3]

[1] See C. K. Hobson, *The Export of Capital*, London, 1914, p. 152, and W. C. Mitchell, *Business Cycles*, pp. 64 and 68–70.

See also the speech of Mr. Felix Schuster, the Governor of the Union Bank of London, Limited, at the meeting in July, 1899. He estimated the total value of these repurchases at £80,000,000. *Bankers' Magazine*, August, 1899, p. 271. See, however, F. W. Taussig, *op. cit.*, p. 220.

[2] *Bankers' Magazine*, December, 1900, p. 687. The United States took half of the £10,000,000 issue of Treasury Bonds floated in August, 1900. *Ibid.*, January, 1901, p. 84. One of the reasons for the flotation in New York, according to the *Economist*, was a desire to secure gold in England, as the stock at the Bank was quite low, August 11, 1900, p. 1142. See also the speech of Mr. Schuster at the meeting of the Union Bank of London in January, 1901, *Bankers' Magazine*, February, 1901, p. 323.

These securities were subsequently resold to England apparently. See the speech of Mr. Felix Schuster at the meeting of the Union Bank of London, Limited, in January, 1903, *Bankers' Magazine*, March, 1903, p. 488. The underlying reason for this was probably that advanced by Mr. Luke Hansard: "Foreign banks take freely English Treasury bills if the rate is attractive, but exchequer bonds or Consols are not taken so readily. The reason is partly that the Treasury bills are generally for short periods, but chiefly because no income-tax is deducted from the discount, whilst even with an affidavit of foreign ownership, the allowance of income-tax cannot be obtained on exchequer bonds or Consols." He said that Americans did not know this when they purchased the exchequer bonds spoken of above. "The Efficiency of Reserves," a paper read before the Manchester and District Bankers' Institute, January 18, 1901, reported in *Bankers' Magazine*, March, 1901, p. 415.

A German Treasury loan of £4,000,000 was also floated in New York. *Economist*, September 22, 1900, p. 1329.

[3] "In the case of Paris, a very large total, which has been estimated at £15,000,000 (and I have even seen larger figures stated in some of the financial journals, such as £20–30,000,000) was taken off this market, thus affording considerable relief, although later on, when these arbitrage transactions have to be settled, these operations may produce an adverse exchange." Hansard, *op. cit.*, p. 412. At the meeting of the Union Bank of London in January, 1901, Mr. Schuster said that he was afraid

The security market in London was upset in May, 1901, by the refusal of some of the joint stock banks to lend after the collapse of the American stock market following the corner in Northern Pacific stock.[1] From the viewpoint of the London market this was only one of the minor circumstances which compelled the market to borrow at the Bank of England. It had a momentary effect of making American securities unpopular in England, and led to investment in home securities rather than foreign.[2]

---

of the increased power of drawing on English gold reserves which was given to foreign nations by these transactions. See *Bankers' Magazine*, February, 1901, pp. 323–324.

These fears were held to be unwarranted by Mr. Walter Leaf, at the meeting of the London and Westminster Bank, in July, 1901. He maintained that there had been no decrease in the holding of American securities by Britishers, and adduced figures of loans made by his bank on such securities as collateral in proof. He could find no significant change in the amount of such loans. *Ibid.*, September, 1901, p. 406.

[1] *Economist*, May 18, 1901, p. 739, and *Bankers' Magazine*, June, 1901, p. 897. See also Mitchell, *Business Cycles*, p. 65.

[2] *Bankers' Magazine*, July, 1901, p. 38.

CHAPTER VII

THE LONDON MONEY MARKET FROM 1902 TO 1913

(1) In the period between the Boer War and 1913, the problems encountered by the Bank of England in controlling the money market became more complex. The preeminent position of England in export markets was threatened by Germany and the United States, and the pressure exerted at times by movements of foreign balances became important. On the other hand, the position of the Bank in relation to the market was strengthened by the further progress of the amalgamation movement, and by the establishments of new relations between the Bank and the joint stock banks. The accumulation of gold stocks by some of the banks, however, placed them in a more independent position, and threatened the position of the Bank of England to some extent.

The recession beginning in 1900 continued until the middle of 1904, when a revival began which lasted until 1907.[1] The lowest point marked by the Thomas index is the third quarter of 1904, and expansion commenced thereafter, and continued steadily until the middle of 1907.[2] Thorp designates 1904 as "depression; revival," and sets the beginning of revival during the last quarter of the year.[3] The Sauerbeck index of prices also turned upward during the same period.[4] The crisis of 1907 came in the late summer, but the succeeding depression was short-lived. It reached its lowest ebb in the last quarter of 1908, and revival began during 1909.[5] There was a momentary halt during 1911, occasioned chiefly by the important strikes of that year, but no real decline began until the latter part of 1913.[6]

During the years of expansion from 1905 to 1907 there was an important drain of gold from the Bank into circulation. During

---

[1] See Table I and Chart XIX, where the series plotted in Charts XVII and XVIII are continued for the period 1902 to 1913. See also Thorp, *op. cit.*, pp. 173–174.
[2] See Table I and Chart XIX.
[3] *Op. cit.*, p. 174.
[4] See Table I and Chart I.
[5] See Table I and Thorp, *op. cit.*, p. 175.
[6] See Table I and *ibid.*, pp. 175–176.

the depression of 1908 there was a small reflux, but the heavy internal flow began again in the latter part of 1908. Except for a lull in 1912, the internal drain was heavy until 1914.[1]

During 1904 and the early part of 1905 there was a marked influx of gold from abroad, and the Bank stock was large in the spring of 1905. During the summer of that year, however, a large foreign drain commenced, and the Bank was forced to take action.[2] For 1905 there was little net movement at the Bank, but in the two following years the flow was heavily inward. The stock of bullion, however, declined somewhat,[3] and the imported gold clearly went into circulation or into the hands of banks.

(2) Considering the lack of gold imports from abroad, at least on the scale of the previous years, and the failure of coin to return in large quantities from the provinces, it was only to be expected that the Bank would attempt to retain control over the money market throughout 1901 and 1902. The heavy issues of Government securities aided, since they helped to drain the market of funds. Moreover, the new Parliamentary powers granted to the Chancellor of the Exchequer in September, 1902, enabled him to borrow in the market rather than at the Bank to meet deficiency advances if he chose.[4] When he chose the former course the funds were taken from the market in the first place, and the control of the Bank was increased to that extent.

As a result of these various factors the Bank was able to temper the course of the market for most of these two years.[5] It did not

---

[1] See Table VI and Chart XIX.

[2] See Table II. Cf. *Economist*, Commercial History and Review of 1905, p. 7.
The *Bankers' Magazine* commented on the flow of gold to New York as follows: "An even stronger influence, however, operated at a later period of the month, when it was found that, owing to the monetary stringency in New York and the low level to which the surplus reserves had fallen there, gold was being bought for the United States. Considerable amounts were purchased in the open market on New York account, but at no time did these shipments show any profit as an exchange operation, and manipulative tactics in Wall Street were probably mainly responsible for them." October, 1905, p. 481.

[3] See Table IX and Chart XIX.

[4] *Ibid.*, October, 1902, p. 493, and January, 1903, p. 98.

[5] The following is suggestive: "We believe that there has never been a period when the Money Market has been dependent on the Bank for so long a time as on the present occasion. Ever since the end of last year, it may be said that borrowers

CHART XIX

GOLD AND CURRENCY MOVEMENTS, VARIATIONS IN THE BULLION STOCK OF
THE BANK, DISCOUNT RATES, AND FLUCTUATIONS
OF BUSINESS, 1902–1913

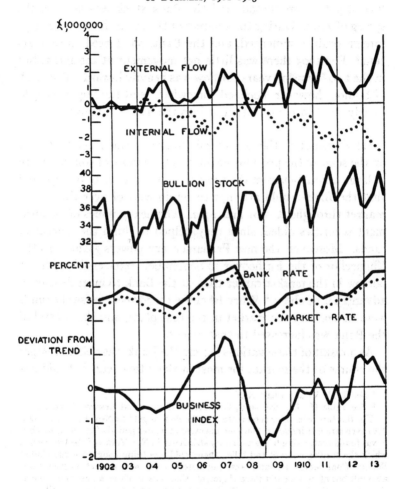

attempt to abuse the power and charge high rates on advances, and there was no real stringency at any time. At the end of each quarter borrowings at the Bank were heavy, and there was considerable lending by that institution between the usual days of stress at the ends of quarters. While imports of gold were not heavy, the stock of bullion at the Bank was fairly high, and there was no need to inflict high rates of interest. Furthermore, in the interest of the Government, which was still doing much borrowing, such rates were inadvisable.

During 1903 the Bank was forced to take action to protect its bullion stock. Conditions elsewhere in the world, particularly in the United States, were calculated to bring a demand for gold from England. The Bank raised its rate several times, and endeavored to make it effective. In addition it raised its buying price for gold bars to increase its holdings when its rate did not succeed in bringing gold to it.[1] But in general it was not as successful in its control as in the years immediately preceding.[2]

have never been able to wholly cancel their engagements, although it is true that much of the indebtedness during recent months has been incurred, not by Lombard Street, but by the Stock Exchange, in connection with 'contangoing' operations on Consols." *Ibid.*, June, 1902, p. 894. See also *ibid.*, January, 1902, p. 84; February, 1902, p. 247; and September, 1902, p. 382. And see Mitchell, *Business Cycles*, p. 67, where the recall of funds borrowed by American speculators in 1900 is mentioned.

In November, 1902, perturbations in the money market resulted from the paying off of British shareholders by the new American shipping combine. A period of easy money was followed by an almost equally severe stringency, and borrowing at the Bank was exceptionally heavy at the end of 1902. The stringency was due in part to the issue of £7,500,000 of Treasury bills in the market under the Chancellor's new power. See *Banker's Magazine*, January, 1903, p. 99, and February, 1903, p. 248.

[1] The *Bankers' Magazine* commented as follows on the lowering of the Bank rate from 4 per cent to 3½ per cent on May 21, 1903: "It has been palpable to close observers of the Money Market for some time past that we have attracted gold, not because of the 4 per cent., but from the action of the Bank in paying a slightly higher price for the metal in the open market, and, although it is quite probable that further withdrawals of gold may take place for Argentina, such withdrawals would probably be made were the Bank rate 3½ or 4 per cent." June, 1903, p. 865.

In November it was noted that the borrowing operations of the Bank were only partially successful, and were then abandoned. "The directors then, on the urgent representation of large financial interests, agreed to do what others would do in circumstances requiring the replenishment of gold stocks; they went into the open market and bought gold at the market price." *Ibid.*, November, 1903, p. 595.

[2] Commenting on the borrowing operations of the Bank in its endeavor to enforce its rate on the market, the *Bankers' Magazine* spoke as follows: "Indeed, it may per-

It has been found previously that in time of depression the country banks tended to reap large profits in comparison with city banks, for their rates on loans and overdrafts did not tend to fall, while the deposit rate, although not as sensitive to Bank rate as in London, declined. But a new feature was added in this regard during the years following 1902. It appears that many municipal corporations had been borrowing largely from banks and the London money market. The proceeds had been spent in many cases in unprofitable ventures on the part of the local governments, and it became increasingly difficult for them to secure loans in the normal way. As a result they began to offer interest on funds deposited with them. The country banks and branches of London banks were consequently forced to keep their deposit rates up in order to compete with these municipal corporations.[1]

(3) Money rates were relatively low during 1904 and the first half of 1905, but during the second half of the latter year rates were advanced, and the Bank rate was at 4 per cent at the end.[2] There was an increasing demand for loans in the country as trade expanded, and balances were withdrawn from London to finance

haps fairly be said that the reason for the average value of money being rather higher than in the previous year is a little artificial, in the sense that the high Bank rate was constantly maintained by this artificial process of borrowing." January, 1904, p. 82.

The Chairman of Lloyds Bank, Limited, said that the chief feature of 1903 was the stagnation in the stock exchange. *Ibid.*, March, 1904, p. 445. It was generally recognized that the funds transferred from the provinces were used in the stock exchange, and the little amount of business there would afford some slight evidence that the amounts transferred from the provinces were small.

[1] In 1904 it was reported that country bank profits were less than in 1903, owing to the fact that the competition of municipal bodies had forced the banks to maintain deposit rates at the former level, while the cheap money rates of some years back had reduced the general lending rate in the provinces from 5 per cent to 4 per cent, where it tended to remain. *Ibid.*, July, 1904, p. 16.

In its review of 1904 the *Bankers' Magazine* was more explicit, and said that rates on deposits in the country were from 2½ to 3 per cent. City banks were allowing around 1 per cent at the same time. January, 1905, p. 49. Cf. *ibid.*, July, 1905, p. 3.

See also the speeches of the Chairmen of Lloyds Bank, Limited; the London City and Midland Bank, Limited; and the Governor of the Union of London and Smiths Bank, Limited, at the meetings in January, 1904, *ibid.*, March, 1904, pp. 447, 459, and 476; and the speech of the Chairman of the London City and Midland Bank, Limited, in January, 1905, *ibid.*, March, 1905, p. 461.

[2] *Bankers' Magazine*, January, 1905, p. 48, and August, 1905, p. 166.

the industrial expansion outside.[1] Despite the increased demand
for funds, it was necessary for the Bank to borrow heavily in October to make its rate effective, and stop the loss of gold [2] In September it had raised its rate to 4 per cent, and this was continued
in force throughout the rest of the year. In December the rate on
advances was raised to 5 per cent.[3] The situation at the time was
precarious due to the paying off of some £14,000,000 of Treasury
Bonds, and market rates fell off to as low as 3 per cent for fine
bills. The Bank's reserve was endangered by the consequent fall
of the French exchange, and the Bank took a new departure and
asked the principal clearing banks to co-operate with it by taking
surplus money off the market, and increasing their deposits at the
Bank thereby.[4] This method was followed again in January,
1906,[5] and the Bank rate was made effective, bringing up market

[1] *Ibid.*, January, 1906, p. 46, and July, 1906, p. 3. It was reported that banks
doing a large country business were compelled to limit their accommodations granted
in London.

[2] *Ibid.*, November, 1905, p. 611; December, 1905, p. 720; and January, 1906, p.
84. Cf. *Economist*, Commercial History and Review of 1905, p. 7.

[3] *Bankers' Magazine*, January, 1906, p. 78.

[4] See *ibid.*, March, 1906, pp. 356–357. *Economist*, Commercial History and Review of 1905, p. 7, and the speech of the Chairman of Lloyds Bank, Limited, in
January, 1906, *Bankers' Magazine*, March, 1906, p. 462.
The changed procedure of the Bank in borrowing from the market was remarked
upon by the *Economist*. It pointed out that the Bank might sell Consols, borrow from
one or two banks, or borrow from all the clearing banks. February 10, 1906, p. 211.
The action of the Bank during this period was severely criticized by the *Economist*. It pointed out that the action of the Bank in attempting to make the 4 per cent
rate effective rather than raising its rate, caused it to advance its loaning rate to 4½
per cent. As a consequence, bills were discounted with the Bank in preference to the
securing of advances. For the moment market rates could be allowed to fall, since
immediate repayment was not required. In the case of advances it was necessary to
make repayments within a shorter time. Moreover, the Bank knew that the issue of
Exchequer Bonds was due on the 7th of December, but it did nothing to absorb
market funds in anticipation of the repayment. Not until a week later did it begin to
operate in the open market, and then only because the rate at the Reichsbank was
raised. Consequently, the market had false leads during the period, and could not be
expected to keep rates up. *Economist*, February 10, 1906, p. 210. Cf. *Bankers' Magazine*, March, 1906, p. 357.

[5] The policy of borrowing from the joint stock banks was continued in the first
quarter of the new year, when the gathering of the revenue normally depletes the
market funds, and this met with much criticism of the policy. See *Economist*, February 3, 1906, p. 165, and February 24, 1906, p. 301. Cf. *Bankers' Magazine*, February, 1906, p. 258, and March, 1906, p. 358.
In the first part of 1907 the Bank also borrowed in the market, and the *Economist*
again criticized it for its action. January 26, 1907, p. 125.

rates and raising the French exchange from the gold export point.[1]

During 1906 the Bank reserve fell until it reached the danger point in the fall, and the Bank was forced to take drastic action.[2] Although the Bank had begun to raise rates during the latter part of 1905, the internal drain was so severe that it was impossible to retain a large stock of bullion and coin at the Bank. It was pointed out at the time that a rise in the Bank rate should have occurred during the late summer of 1906. The United States had been borrowing large funds in England by the drawing of finance bills,[3] but, due to the dilatory action of the Bank, these bills were renewed in the early fall instead of being paid off.[4] And the Bank was finally forced to act in a more severe manner than would otherwise have been the case.[5] Its rate was advanced to 5 and then to 6 per cent in September and October, which created a

---

[1] The difficulties of this period were engendered chiefly by the controversy between France and Germany over Morocco. There was a tendency for France to withdraw balances employed in England, and exports of gold during the last half of the year were large from the Bank of England. See, e.g., the speeches of the Chairmen of the London City and Midland Bank, Limited; Lloyds Bank, Limited; and the Governor of the Union of London and Smiths Bank, Limited, in January, 1906, *Bankers' Magazine*, March, 1906, pp. 458, 463, and 473.

[2] In the *Economist* of June 9, 1906, p. 957, it is reported that the Bank had hinted to the market that interest rates should not be allowed to fall much, or they would drain funds from Lombard Street. No important action followed, however, and the situation became grave in the fall.

[3] See *Bankers' Magazine*, October, 1906, p. 504.

[4] In August the rate was lowered to 4 per cent for discounts and 4½ per cent for loans, and it continued to loan to its own customers at 4 per cent and to the market at 4½ per cent in September. Moreover, the Bank kept its rate on advances ½ per cent above the rate on discounts, and the market sold bills to it instead of borrowing on collateral when it needed funds. Since immediate repayment was unnecessary, market rates were allowed to drop. *Ibid.*, November, 1906, pp. 565–566 and 605–607. Cf. *Economist*, Commercial History and Review of 1906, p. 6.

[5] Not only did the Bank advance its rate, but it intimated to the banks holding American acceptances (finance paper) that such bills were a menace to the stability of the London market. Their repayment was demanded at maturity as a result, and the liquidation of these bills from December, 1906, was a very important factor. See O. M. W. Sprague, *History of Crises Under the National Banking System*, (United States) National Monetary Commission, 61st Congress, 2nd Session, Senate Document No. 538, p. 241.

The *Economist* pertinently remarked that when the joint stock banks facilitated the withdrawal of gold by discounting large amounts of finance bills for America they should assist in maintaining gold reserves from which exports are drawn. September 1, 1906, p. 1429.

scare on the money market. Continental rates were also high at the time, and relatively little gold was secured from abroad until late in the year. It was feared that the Bank rate would be raised to 7 per cent, since little specie was acquired under the 6 per cent rate, and Continental countries, particularly France, began to release gold for export.[1] This situation continued into 1907, with America hesitant about drawing gold, knowing that it would mean a higher Bank rate.[2]

The English crisis was hastened by the outbreak of the severe American panic in the late fall of 1907.[3] At the time the Bank of England had control over the London market,[4] and it speedily ad-

---

[1] It was generally believed that the Bank of France had offered a loan to the Bank of England, but that the latter had refused it, preferring that gold should enter in the usual manner. *Bankers' Magazine*, December, 1906, pp. 725–726. It was at this time that the Bank of France began the practice of purchasing sterling bills to aid the London market. Instead of raising their rate of discount, they facilitated the flow of gold to England by buying bills. Liesse maintains that this expedient was more "economic" than the archaic method of making discreet loans to the Bank of England. See André Liesse, *The Evolution of Credit and Banks in France*, (United States) National Monetary Commission, 61st Congress, 2nd Session, Senate Document No. 522, p. 230. The total amount involved in 1906 was 75,000,000 francs, according to Maurice Patron, *The Bank of France in Its Relation to National and International Credit*, (United States) National Monetary Commission, 61st Congress, 2nd Session, Senate Document No. 494, p. 143. See also White, *op. cit.*, p. 195.

The Bank of France withdrew gold from England in the spring of 1907 by letting the bills held run off as they matured. See *Economist*, April 20, 1907, p. 673.

[2] "This fact, however, was clearly recognized by American bankers, and although at one time some gold had nearly been engaged from the open market the effort was abandoned, ill effects which would arise from an advance in the Bank Rate here being universally admitted . . . and the New York exchange being in the meantime greatly influenced by the payment on the part of Paris and London for several millions sterling of United States Railroad Notes placed at those centres, the cable transfer rate on London stood for days at a point showing large profits on shipments of bar gold from the market in London to New York." *Bankers' Magazine*, April, 1907, p. 586. The situation was made acute by the stock market panic in Wall Street in March. Relief was afforded by the Secretary of the Treasury under powers granted by the Aldrich Act, and the rate of exchange on London fell. "So sharply, indeed, did the rate move in favour of London, that a large amount of gold which had actually been purchased here on New York account was resold, and from that date (April 3) the tendency of money rates in Lombard Street may be said to have completely changed." *Ibid.*, May, 1907, p. 739. See also *Economist*, March 16, 1907, p. 449, and April 6, 1907, p. 577.

[3] The movements of the crisis period have been dealt with by many writers, and it is unnecessary to go into detail concerning them. See Mitchell, *Business Cycles*, pp. 540–543, and Sprague, *op. cit.*, pp. 276–285.

[4] See *Bankers' Magazine*, November, 1907, p. 619.

vanced its rate to 6, and then to 7 per cent, and market rates rose
with it. The situation was more difficult than in 1906, largely
because few American finance bills were held in London at the
time. The admonitions of the Bank of England in the fall of the
preceding year kept the discounting of such paper to a minimum
in 1907.[1] Consequently, no gold had been drawn to the United
States earlier in the year,[2] and the full weight of the American demands for gold were concentrated in the weeks following the outbreak of the panic. From the standpoint of the United States this
was advantageous, for the receipts from crop shipments did not
have to be applied to the repayment of finance bills, and the heavy
remittances necessitated brought gold from England.[3] Large
shipments of gold to New York during November and December
occurred when the exchange in New York on London was actually
above the gold *export* point of normal times. The high rate of exchange was due to the premium on gold in the United States,[4] and
the effective rate was below the gold import point when allowance was made for the premium.

The immediate effect of the American panic was to draw specie
from England; but the advance of the Bank rate to 7 per cent
brought gold to London from other countries, and England became a clearing center through which gold passed from the rest of
the world to the United States. The Bank of England lost little
gold as a consequence: the net export for the fourth quarter of
1907 amounted to less than one million pounds, and a net gain
accrued to the Bank in the first three quarters of the year.[5] This
achievement was heralded as the acme of success of the English
banking system under the Act of 1844.[6] But in fact, the Bank
rate of 7 per cent drew gold only because of the action of the Bank
of France and the Reichsbank in permitting specie to be drawn
from their hoards.[7] When account is taken of the sums drawn
from these two countries, the amount unearmarked from the Indian account at the Bank of England, and the newly arriving gold

---

[1] See Sprague, *op. cit.*, p. 246.
[2] *Ibid.*, p. 245.
[3] *Ibid.*, pp. 283 and 285.
[4] *Ibid.*, p. 283.
[5] See Table II.
[6] See, e.g., *Interviews*, pp. 26–27.
[7] See White, *op. cit.*, pp. 195–196.

from mining countries, the receipts from other countries are found to be very small.[1] In other words, the high Bank rate brought gold merely because the two Continental banks preferred to aid the Bank of England, rather than suffer higher discount rates in their own countries.[2] Otherwise, the English Bank rate might have had to go to untold heights to keep their Reserve up, and the trade of the country would have suffered severely.

(4) Immediately the panic in the United States was ended and cash payments renewed, the Bank rate was lowered and conditions in London became easier. Coin began to return from the provinces in 1908.[3] The flow of cash into the Bank from the in-

---

[1] This fact was first pointed out, so far as I have found, in a letter to the Editor of the *Bankers' Magazine*, June, 1908, pp. 874–875, by Arthur Lee. He says that of the total shipments to America only £367,422 was drawn from countries other than France or Germany, with the exception of small amounts from Russia and Belgium. The total from the Continental countries named was £9,817,151; it came in response to special appeals for aid, and not because of the high discount rates in London. In addition some £6,613,404 came from mining countries; and a total of £17,437,077 was sent to the United States in November and December.

An examination of the English statistics substantiates Lee's results in general. It is difficult to determine the source of all the receipts of specie by the Bank, but the amounts received from the Scandinavian countries, Turkey, and South America are larger than he gives. Moreover, the receipts from Australia were substantial, and he has evidently treated them as supplies from mining countries. Part of these shipments were in sovereigns, however, and it is possible that they were not commodity specie shipments at all, although one cannot be certain, since the large movement from Australia was in that form after the establishment of the mint at Freemantle. In his calculation Lee includes the two months of November and December, but by the middle of the latter month the heavy flow of gold had ceased, and withdrawals of gold from England for South America had begun again. Certainly it can be said that the high rates were of little aid in the last two weeks of December. The corrections necessary make little change in his figures, and his point is well taken.

See also Hartley Withers, *The Meaning of Money*, 3rd ed. London, 1909, pp. 102 and 221.

[2] It was pointed out by the Chairman of the London City and Midland Bank, Limited, that the high rates imposed by the Bank of England meant that English banks could not discount American finance bills, and that ultimately France became the real international banker. *Bankers' Magazine*, March, 1908, p. 469. Cf. also Liesse, *op. cit.*, p. 230, and Patron, *op. cit.*, p. 144. The latter points out that such movements of bills and gold were a definite part of the financial policy of the Bank of France, and that it was the great gold reservoir of the world (*ibid.*, pp. 112 and 143).

[3] This reflux was commented upon by the *Bankers' Magazine* in February, 1908, p. 286: "Our own Bank Returns have shown that cash is returning from the provinces with greater rapidity than during recent years, and in Germany and the United States the same indications are apparent." See also *ibid.*, June, 1908, p. 859, and October, 1908, p. 607. And see Table VI.

ternal circulation was large in the first quarter of the year, and the outflow from the Bank in the second quarter was about as usual. In the third quarter the usual efflux was reversed, and the Bank gained gold from the country circulation. The influx was short-lived, however, and the Bank showed only a small gain for the year. By 1909 the heavy flow of gold into circulation was again apparent.[1] It is probable that the joint stock banks intercepted much of the returning coin and kept it in their own vaults during this year, for they had begun to accumulate large sums irrespective of the holdings of the Bank of England.[2] This factor helps to explain the small gains of the Bank of England during the year. It is apparent again during the period of expansion and crisis that the Bank gained gold during prosperity from other countries, but that its new supplies were taken by the country to add to the circulation. It is also clear that the reflux to the Bank began shortly after the crisis. Due to the action of some of the joint stock banks, the Bank of England did not receive as much of the return flow as in earlier cycles, but it is significant that the reflux began more quickly than in other cases.

All during 1908, aside from the first quarter, the Bank tended to lose gold on foreign account. Much of the gold sent to the United States in the preceding fall was returned directly to Europe; but England loaned large sums to Continental countries, and it was unable to acquire the gold arriving weekly from South Africa as a consequence.[3] For a period of nearly a year from April, 1908, the Bank did not purchase any bar gold in the market.[4] This heavy movement of gold to the Continent ceased with 1909, and the position of the Bank was improved. The exports from England were made possible by the market arrivals, and by the return of coin from the country, in addition to the unearmarking of some £2,000,000 more from the Indian balances at the Bank.[5] But the

---

[1] See Table VI.
[2] See above, pp. 88–91. In August it was reported that "The Market was, undoubtedly, assisted by the large amounts of gold which had returned from the provinces during the earlier part of the year, while the joint stock banks, which are understood to have been keeping larger daily reserves, were calling in cash less freely from the market for monthly balance sheet purposes." *Ibid.*, August, 1908, p. 147.
[3] *Ibid.*, July, 1908, p. 39, and September, 1908, p. 319.
[4] *Ibid.*, March, 1909, p. 396.   [5] *Ibid.*, January, 1909, p. 85.

Bank stock did not decline, and tended to increase slightly during 1908, despite the European demands.

During 1908 the American crisis showed its effects in Europe. The resulting trade depression in England affected the operations of banks, and a decline in profits was suffered, particularly by the London banks.[1] Those in the provinces were able to keep their earnings up longer, as usual, and banks operating in both sections found that business in the latter regions was of substantial aid.[2] The half-yearly dividend period at the end of June was passed easily, and there was much less borrowing from the Bank than in previous years.

(5) During the early part of 1909 there was a wide difference between the Bank rate and market rates of discount.[3] It was necessary for the Bank to take action later in the fall as a consequence, and it raised its rate to 5 per cent in October and began to borrow in the market to make it effective. The immediate cause for the advance was the large export of specie due to very heavy lending to foreign countries, especially the United States.[4] The Bank also raised its buying price for gold bars, but its operations were only partially successful.[5] It fared better in the following year, however, and the imports during the third quarter of 1910 aggregated over £11,000,000. The autumnal demands took a considerable part of the new gold, and the Bank was forced to take action. The Bank was tardy in checking the fall of money rates in the first half of the year, and it was called upon to take severe action in the latter half of the year.[6] In 1911 the Bank raised the

---

[1] The *Bankers' Magazine* in its review of 1908 remarked that the stock market was better for loans in that year, as usual when trade was depressed. On the other hand deposit rates in the country did not fall as much as in the city of London (2½ to 3 per cent as compared to 1 per cent in London), which would tend to retain money in the provinces. Banks with a minimum charge for loans fared better than other banks at this time. This advantage was felt most at the beginning of a period of depression and cheap money, as competition in the end affected all banking business. *Ibid.*, January, 1909, pp. 38–39. Cf. the speeches of the Chairmen of the London and Southwestern Bank, Limited, and the London and Provincial Bank, Limited, at the annual meetings in January, 1909, *ibid.*, February, 1909, p. 188.

[2] *Ibid.*, August, 1908, pp. 117–118.   [3] *Ibid.*, July, 1909, p. 1.

[4] *Ibid.*, November, 1909, pp. 660–661.   [5] *Ibid.*, December, 1909, p. 774.

[6] See *ibid.*, January, 1910, p. 72, and January, 1911, p. 74.

rate in March, which was only the second time in the history of the the Bank that the rate had risen during this month.[1] Despite intermittent losses, the Bank tended to gain gold from abroad during these years from 1909 to 1913.[2] Its problem was made more difficult by the large accumulations of gold which were being made by the central banks of the Continent, particularly Germany and Russia.

Nevertheless, the Bank did gain some gold for its stock, and the gain would have been great if it had not been for two internal factors. The first of these was the large withdrawal of gold into internal circulation and the arts on account of the expansion of trade and industry.[3] This amount was large in all years up to the beginning of the War in 1914, and contributed greatly to the depletion of the Bank stock. The second factor was the large accumulation of gold in the vaults of the large joint stock banks.[4] This gold was secured either from incoming supplies, which might otherwise have gone into the Bank store, or by intercepting the normal seasonal return of coin to the Bank.

One other aspect of the international situation is of interest. At several times during this period gold imports into England were induced, not by the actual rates imposed by the Bank, but by the fear of higher rates that might be imposed. It will be recalled that in the fall of 1906 and of 1907 the Bank of France began to purchase sterling bills and to pay for them in gold, for fear that the Bank of England rate would be raised to 7 and 8 per cent, respectively.[5] The same procedure was adopted again in 1910 and in 1913. In these latter cases there appears to have been an

---

[1] *Ibid.*, November, 1910, p. 637.

[2] See Table II and Chart XIX.

[3] See Table VI. It was generally recognized that this drain was proceeding. See, e.g., the speech of the Governor of the Union of London and Smiths Bank, Limited, at the meeting in January, 1910; *Bankers' Magazine*, March, 1910, p. 477; and see *ibid.*, July, 1911, p. 3; January, 1912, p. 32; April, 1912, p. 613; August, 1912, p. 177; and January, 1913, p. 31. It was reported that the coal strike in 1912 caused the banks to take large amounts of coin to the provinces to prepare for emergencies. The following quotation reprinted by the *Bankers' Magazine* from the *Standard* is typical: "The provinces have been drawing upon the Bank of England largely from time to time for sovereigns, and these sovereigns return only to a small extent from circulation." July, 1913, p. 60.

[4] See above, pp. 88–91.   [5] See above, pp. 143–145.

agreement made at the time that the bills would be repaid with *gold* on reaching maturity.[1] In other words, the Bank of France was willing to let gold go temporarily to avoid any disturbance in the French market, but they were not willing to lose the specie permanently. It seems clear that they were to that extent becoming the central bank for England as well as France, and were permitting the use of their bullion stock because the English system was not competent or willing to take care of seasonal and other needs.

(6) Two important developments occurred in London during the period after 1901. The first was the general agreement on the part of London joint stock banks not to loan day-to-day money at less than the deposit rate of interest.[2] This was an important change, for it removed some of the extreme fluctuations in the London money rates, and gave greater firmness to the money market.

The second innovation was the commencement of conferences between the joint stock banks of London and the Bank of England. It was agreed that one meeting in each quarter of the Committee of the London Clearing Bankers would be held at the Bank of England.[3] This was an important development as it meant closer co-operation would be evidenced between the Bank and the important joint stock banks.

(7) During 1912 and 1913 the stringency in the London market was severe, and resort to the Bank was made throughout most of the months. The trade activity, with resultant flows of cash and of loanable funds — which would otherwise have been lent in the London market — to the provinces was the most important factor; but the accumulation of especially large Government balances in 1912 also contributed. In the latter case, the large

---

[1] *Ibid.*, December, 1910, p. 755; January, 1911, p. 93.

[2] *Ibid.*, January, 1911, p. 30.

[3] The joint stock banks at the meeting on July 20 unanimously adopted the resolution asking the Bank to join in their meetings. The Governor of the Bank replied at the annual Mansion House Dinner, saying that the Bank was very glad to have the proposal of the bankers. He also referred to the real unity of interests between them and the Bank. *Ibid.*, August, 1911, p. 152, and September, 1911, p. 406.

surplus was not expended in purchasing Consols for the Sinking Fund, but was held at the Bank on deposit.[1] The Chancellor was subjected to much criticism for his policy in this regard. The market was forced to go to the Bank for funds which would normally have been supplied by the release of Exchequer money. There was no such locking up of Government money in 1913, but by that time the political difficulties on the Continent had invoked a spirit of caution into the market, which had been lacking until late in the preceding year. The Bank had raised its rate to 5 per cent in October, 1912, and it was not changed until April of the following year.[2] But rates continued high in 1913 on the average, and the Bank of England had a fair amount of control taking the year as a whole.[3] The bullion stock remained at a reasonable level, however, due to imports of specie, partly through the purchase of the newly arriving gold, and partly because of receipts from South America.[4]

During the period from the Boer War to 1913, the increase in gold reserves of the larger joint stock banks gave them somewhat greater freedom from the Bank of England. To the extent that they held excess reserves of gold during ordinary times, they had become small central banks. But it is apparent that in periods of stress, at least, they were more willing to co-operate with the Bank of England in bringing the money market under control than they had been earlier.

The more important developments in the relations of the Bank of England to the money market and the banking system are considered in the following chapter. The changes have been mentioned in the chapters on the money market, but they are brought together for purposes of analysis.

[1] See *ibid.*, November, 1912, p. 625, and January, 1913, p. 99. The surplus amounted to some £6,500,000, and it was carried to a "suspense" account, and not released for debt redemption.

[2] *Ibid.*, May, 1913, p. 755.

[3] *Ibid.*, January, 1914, pp. 31 and 85–88. The Bank rate did not fall below 4½ per cent all the year, and it was at 5 per cent for much of the time.

[4] *Ibid.*, pp. 94–95 and 103–104.

CHAPTER VIII

RELATIONS OF THE BANK OF ENGLAND TO THE MONEY MARKET AND THE BANKING SYSTEM

(1) In the middle of the nineteenth century the Bank of England was the leader in the money market principally because it always supplied a portion of the supply of funds seeking investment. Sykes quotes the *Bankers' Magazine* of 1863 as follows: "Bankers merely look to the Bank of England returns, and tacitly follow in the wake of that establishment." [1] He goes on to say: "The Bank of England was then so strong that it could have absorbed all the other London banks, their capitals and their reserves, and yet its own capital would not have been exhausted." [2] And he calls attention to Bagehot's well-known quotation that: "At all ordinary times there is not money enough in Lombard Street to discount all the bills in London without taking some money from the Bank of England." [3] And Feavearyear maintains that at times the competition of the Bank for bills was responsible for the forcing down of market rates. Except when the Bank had to protect or add to its Reserve, Bank rate and market rate were held close together by this competition. [4]

But after 1857, and especially after 1866, the policy of the Bank with respect to discounts was gradually altered. This change came in part because of the growth of the London joint stock banks, and the increase in the capital available for lending in the hands of private banks and discount houses, and in part because of the lessons learned in the crises of 1847, 1857, and 1866, and the repeated strictures of Bagehot on the necessity of keeping adequate reserves for any emergency. In 1878 the Bank definitely accepted the new situation when it announced that it would no longer lend to bill brokers at the market rate, but would restrict

---

[1] *Op. cit.*, p. 169.   [2] *Loc. cit.*

[3] Bagehot, *op. cit.*, p. 114.

[4] Feavearyear, *op. cit.*, p. 282. See also Marshall, *op. cit.*, pp. 127–128, and Gregory, *op. cit.*, pp. xxxiii–xxxiv.

discounts at that rate to its regular customers only, when market rate was below Bank rate.[1] The Bank rate consequently became completely ineffective except for certain periods, and it no longer was the governing rate at which the best bills were discounted.[2]

There were generally periods within each year when the rate was effective, and occasionally the influence of the Bank was felt over longer periods. Within each year the market was normally bare of funds when there were heavy payments to be met on account of dividends, especially at the end of each quarter. This phenomenon was especially pronounced in the first quarter of the year, for at that time the collection of taxes swept large sums of cash and deposit credit into the Bank. The same situation sometimes prevailed when internal drains of coin or notes for holiday or harvest requirements were exceptionally large.[3] Any large shipment of gold from the Bank to a foreign nation always brought strong psychological, as well as mechanistic, forces to bear upon the market, and was likely to cause an upward movement of mar-

---

[1] See Palgrave, *op. cit.*, p. 55, and Feavearyear, *op. cit.*, pp. 281–285. Feavearyear holds that Bagehot's statement quoted above conveys the erroneous impression that the Bank's funds were always needed in the market. He states that a reversal of the Bank policy with respect to discounts would bring an increase in the demands for accommodation, and thus make the Bank's funds just as indispensable as in the middle of the century. This position is perfectly sound to the extent that easing of rates by any lender brings an increase in the demand for funds, but unless the lender has very large funds available in comparison with the total demands existing at a given time, the effect of easier lending conditions will be very slight upon the market as a whole. Certainly by 1913 the resources of the Bank of England would not have affected market rates appreciably if it had ceased to be a central bank and had lent to the market upon the same terms as other lenders.

[2] The *Bankers' Magazine* maintained that the Bank rate had become nothing more than a signal that the stock of specie was low. "Except at periods of great stress and demand, the Bank of England is now always practically out of the market as a taker of good commercial or less even of first rate 'bank-bills'." April, 1890, p. 527.

And the *Economist* comments as follows: "This want of power on the part of the Bank to make its rate effective has been increasing year by year, while every year the dependence of the outside market upon the Bank has increased, because our great joint-stock institutions have been gradually diminishing the proportion of their cash reserves to their liabilities." Commercial History and Review of 1890, p. 4. The remedies proposed were the payment of interest on deposits by the Bank to increase its control of funds, the keeping of larger reserves by the joint-stock banks, and agreement among these banks to assist the Bank of England in keeping the Bank rate effective.

[3] Cf. Feavearyear, *op. cit.*, p. 287.

ket rates.[1] Most important of all were needs resulting from the expansion of business in the periods just preceding crises.[2] Only in the last case was it necessary for the market to resort to the Bank for loans over any extended time.

By 1881 the Bank had learned to alter its discount rate in a systematic manner in order to control gold flows; and the state of its Reserve was the criterion upon which discount policy was based.[3] It had accepted tacitly the doctrine that changes in money rates would ultimately affect commodity prices and industrial production, and readjust the international balance of trade. Moreover, it had discovered that rate increases stopped foreign drains of specie and enabled the Bank to maintain a stronger reserve position, although the full significance of this causal con-

[1] As the Bank rate might be expected to rise, bringing deposit rates up, the profit on bills held at the time by banks or discount houses would fall, and they would tend to raise their rates on new bills to compensate for this loss. It was most desirable for the banks to hold bills when discount rates were *high* and *declining*, for lower deposit rates would hold and their total profits would be large.

Even more important was the fact that gold was secured from the Bank and paid for by check on one of the clearing banks. The balance of the latter at the Bank would be drawn down, and the tendency would be for it to call some of its loans, thus decreasing funds in the market, and tend to cause a rise in rates.

[2] Cf. Withers, *op. cit.*, pp. 234–235, and Otto Hulftegger, *Die Bank von England*, Zurich, 1915, pp. 173–174, for somewhat similar classifications.

[3] The state of the discount policy at that time is admirably summed up by B. H. Beckhart, *The Discount Policy of the Federal Reserve System*, New York, 1924, p. 29. He says: "It may be concluded, then, that the directors of the Bank of England had a clear concept, during the crisis of 1857, of the procedure which a central bank should follow in the management of its discount policy — a knowledge which they had mastered sometime after the crisis of 1847 but which in its entirety was the outgrowth of the experience through which the Bank passed from 1797 to 1850. The propositions thus established as fundamental regarding the management of the Bank rates were:

"(1) That excessive issues on the part of a central bank in the form of notes or deposits will raise prices, drive gold abroad and depreciate the foreign exchanges.

"(2) That deflation may be brought about, imports of gold and capital induced, and the exchanges restored, through elevating the Bank rate; and this method of contraction is to be preferred to the rationing of credit.

"(3) That the bank rate should ordinarily be kept above the market rate, and that it should anticipate advances in the market rate in order to curb inflationary tendencies.

"(4) That a central bank should discount liberally but at a high rate during a crisis.

"(5) That a central bank, as the holder of the ultimate banking reserve of a nation, should be managed with a view of protecting this reserve rather than with a view of making profits."

nection was not appreciated. The primary machinery of the discount policy was supplemented later in the century only because of the altered relation of the Bank to the London money market. The price-specie flow mechanism, supplemented by use of the discount rate, was felt to be adequate for the adjustment of the exchanges, the price level, and the international balance of payments.[1]

The lack of power over the market at times, however, compelled the Bank to resort to other means than the discount rate to obtain the necessary control. The policy with respect to gold also changed somewhat, due to the deficiency of the reserves. We shall turn to this latter question before examining the problems of discount rates and open-market operations.

(2) Even in the earliest years of the period studied the Bank raised its *selling* rate for bullion,[2] and its prices for foreign coin were also subject to change. But such changes were not so common as in later years. The first instance which I have found of a change of policy with respect to the *buying* rate for gold bars was in 1890;[3] and during the Baring Crisis period the Bank pursued a definite policy of control in order to secure gold. But from 1892 to 1898 there does not appear to have been any increase in the

[1] That such was the case up to 1914 is indicated by the following statement from the Cunliffe Committee, *First Interim Report*, p. 6: "Under an effective gold standard all export demands for gold must be freely met. A further essential condition of the restoration and maintenance of such a standard is therefore that some machinery shall exist to check foreign drains when they threaten to deplete the gold reserves. The recognised machinery for this purpose is the Bank of England discount rate. Whenever before the war the Bank's reserves were being depleted, the rate of discount was raised. This, as we have already explained, by reacting upon the rates for money generally, acted as a check which operated in two ways. On the one hand, raised money rates tended directly to attract gold to this country or to keep here gold that might have left. On the other hand, by lessening the demands for loans for business purposes, they tended to check expenditure and so to lower prices in this country, with the result that imports were discouraged and exports encouraged, and the exchanges thereby turned in our favour. Unless this two-fold check is kept in working order the whole currency system will be imperilled. To maintain the connection between a gold drain and a rise in the rate of discount is essential to the safety of the reserves."

[2] See above, pp. 44–45, and 109.

[3] The *Economist* of February 15, 1890, p. 197, notes the rise of the buying rate by ½d. It continues: ". . . and it is understood, also, that for the gold that was received the other day from Paris something more than the usual price was paid."

buying rate for gold bars at the Bank. In the latter year, the rate was raised above 77s. 9d.,[1] and from that time on the Bank frequently purchased bullion at prices ranging up to 77s. 10½d.[2]

The Bank instituted a policy of interest-free advances to importers of gold in 1898.[3] This aid was granted very infrequently, however. The Bank also adopted the plan of paying bullion dealers in light coin — within the limits of the Mint tolerance — in order to prevent exports of specie. This device was apparently used for the first time late in the nineteenth century.

A more active policy with respect to gold clearly began in the late nineties.[4] During periods of depression there was little need for such action, but in expansion, when gold was demanded for export, the Bank was definitely aggressive. The limits within which the Bank could determine specie movements were small, but the evidence shows that use was made of the Bank's power to supplement its discount policy.

(3) The securing of funds from the Bank by the market was achieved in two ways — by the discount of bills, and by the advance of credits against collateral security. During the period studied there was a noticeable trend toward the favoring of the former method. Rates on loans were kept above those on discounts, particularly during periods of tight money. This practice was used as early as 1881,[5] but no consistent use of the policy was made until after 1887.[6] Moreover, the discount of bills was favored by increasing the usance of bills acceptable for discount. Until 1894 the limit was fifteen days, but in that year it was raised

----

[1] See above, p. 132.

[2] See *Bankers' Magazine*, and *Economist, passim*. On the limits to which the price could rise, see above, p. 44.

[3] See above, p. 132.

[4] Speaking of the various devices used by the Bank to influence the movement of gold, Lord Swaytheling remarked that the Bank formerly did not do these things. See *Interviews*, p. 98.

[5] See above, p. 109, ff. 1.

[6] In December, 1888, the Bank raised its rate for advances to 6 per cent, ". . . thus following the example of the great continental banks, which are in the habit of charging more for loans than for the discount of bills." *Bankers' Magazine*, January, 1889, p. 41. See also Gregory, *op. cit.*, p. xxxiv.

to thirty days,[1] and in 1897 to ninety days.[2] At any particular time the Bank might refuse to discount long bills, for its policy depended upon its own needs and the conditions existing in the money market. But it is clear that the average usance of bills held increased with the passage of time.[3]

The requirements of the Bank with respect to advances changed somewhat at times. In 1883 loans were granted only in the six weeks preceding the end of the quarterly periods.[4] This rule was relaxed shortly, however. A more important change occurred in 1894, when the closing hour for loans on collateral was altered from 3.30 to 2.30 P.M. The action was taken for the convenience of the Bank, and to protect it from being called upon as the last resort for loans.[5] On the other hand, in 1897 the Bank permitted loans for three days, and did not insist that they be taken for a full week.[6] No evidence as to the continuance of this policy has been found, however.

The discrimination shown by the Bank in favor of discounts may have been due to a desire to divert funds from the stock market into commercial channels, but it is not possible to draw a definite conclusion. It is evident, however, that the Bank found it advantageous to "shock" the money market at times by altering its methods of lending.[7] While such changes could not have lasting effects, they tended to disrupt the working of a highly sensitive credit organization, and gave the central bank more

---

[1] See *ibid.*, March, 1894, p. 395.

[2] See above, p. 131.

[3] The average usance of bills discounted by the Bank was given as forty to fifty days by the Governor and Directors in reply to a question of the National Monetary Commission, 1909. See *Interviews*, pp. 19–20.

[4] See above, p. 110.

[5] The *Bankers' Magazine* comments on this change are as follows: "This regulation was for the convenience of the Bank, but dealers felt they would be shorn of the privilege of going to the Bank in the last resort, after trying the rest of the market for cheaper loans. It seems that they had developed a habit of throwing stocks on the counters of the Bank of England at the last moment, and almost demanding loans on stock." The remedy, according to the *Bankers' Magazine*, was to become a regular customer of the Bank. January, 1895, p. 66.

[6] See above, p. 131.

[7] In the spring of 1890 the Bank even took strong measures at a time when the payment of taxes should have brought action to ease the market. See above, p. 116, n. 4. The episode resembles that in the spring of 1929 in New York.

power. In the main the Bank modified its policy in accordance with changes in conditions in the market.

More important, however, was the extended use of open-market operations. The Bank had begun the practice of selling and purchasing securities much earlier.[1] But apart from some shifting of securities during the thirties, the operations were designed only to smooth the money market over seasonal pressures. In the years following 1881 the Bank resorted more freely to such operations, and to the extent that they were carried on more frequently in the periods of business prosperity when the Reserve was falling, there was some cyclical significance in the operations. I have found no evidence of purchases of securities during depressions to ease the market.

The Bank employed three methods in open-market dealings. The use of one of them is definitely seen in 1885,[2] e.g., when the market was drained of funds by *borrowing* on the part of the Bank through a broker, giving Consols as collateral security.[3] This was the usual practice, but the Bank also operated through the medium of the stock exchange. It borrowed from the dealers by lending Consols from one account to another, which in effect was a sale for cash and a repurchase for the next account, or a shorter time as might be arranged. This was known as borrowing on "contango."[4] At times the Bank also sold Consols outright, but this was practiced very rarely.[5] The use of Treasury bills in open-market operations is primarily a post-war development, because there were rarely more than a few millions of them outstanding at any time before 1914.

In 1905 the Bank secured the co-operation of the large clearing banks of London in draining the market of funds, and this policy was utilized again in 1906. Later the Bank established closer relations with these banks, and conferences began to be held in which

---

[1] See *The Bank Charter Inquiry, 1832, on the Regulation of the Note Issues, Parliamentary Papers*, 1831–1832, vol. VI, Questions 72, 87, 255–256, and 2394–2398, and Gregory, *op. cit.*, vol. I, p. xi.

[2] See above, p. 112.

[3] See *Interviews*, pp. 29 and 51.

[4] Spalding, *op. cit.*, p. 84, and Hartley Withers, *Stocks and Shares*, New York, 1915, pp. 273–282.

[5] *Interviews*, p. 29.

both parties participated.[1] This co-operation was mutually advantageous, for the Bank gained greater control over the money market, and the banks gained through the higher rates imposed as a consequence. Bankers long had objected to the paying of deposit rates above the day-to-day rate for loans. With the concentration of control brought by the conferences with the Bank of England, the joint stock banks were able to refuse to lend at less than the rate of deposits.[2]

In this period from 1881 to 1914 a definite change took place in the relation of the Bank of England to the London money market. At the beginning of the period the Bank had less contact than at any time since the passing of the Act of 1844. But by the outbreak of the war in 1914 the Bank had regained most of its lost power, and was in a position generally to make its rate effective on the money market.[3] The gradual reduction in the spread between the Bank rate and the market rate is evidence of this increased control.[4]

(4) The study of the structure of the English banking rates reveals another mechanism of control which was of particular significance during the earlier years of the period under examination. It is difficult to secure adequate evidence on rates, but it is apparent that the relationship of the Bank rate to deposit rates and charges for loans and advances [5] was very significant.

The customary rate on loans in London and in the country outside varied with the Bank rate of the day. This was particularly important for the country banks, since the proportion of earning assets in the form of loans was higher than in the case of London

[1] See above, p. 150.
[2] See above, p. 150.
[3] Cf. Feavearyear, *op. cit.*, pp. 288 and 292.
[4] See Table XIV and Chart XVI.
[5] Cf. F. E. Steele, *Present-Day Banking: Its Methods, Tendencies and Characteristics*, London, 1909, p. 107.
A full explanation of the various types of loans is given by the *Bankers' Magazine*, January, 1900, pp. 5–9. There were three types: the ordinary overdrawn account, the cash credit account, and the loan account. In the case of the first two, interest was paid at the termination of the loan, or at the end of the bank's financial year. In the case of the loan, interest was deducted in advance. The cash credit was used most largely in Scotland.

160　BRITISH INTERNATIONAL GOLD MOVEMENTS

banks. A high Bank rate was, therefore, regarded as an affliction by the industry dependent upon local banks.[1] Rates on deposits also varied in accordance with the Bank rate.[2] This practice became less prevalent in the provinces as time passed, however, as

[1] The following passages are indicative of the power of the Bank rate in the country:

In 1883 the *Economist* pointed out that the Bank rate is "rigidly adhered to by Scotch banks." The customers of the banks objected, since the Bank rate was not effective in London. Commercial History and Review of 1882, p. 5.

W. J. Atchison shows that when the Bank rate is not in accordance with market rates, it is protecting its Reserve at the expense of the trading community; "For a difference of 1 per cent. in the rate means that whilst its fellow bankers are in a position to lend, the Bank of England is not in that position, and as the Bank rate is the basis for bargains all over the country, its retention at a level of 1 per cent. above the market is, if avoidable, an injury to trade." *Op. cit.*, p. 310.

The *Economist* of February 22, 1890, p. 220, comments as follows: "And seeing how heavily the 6 per cent. rate was pressing upon country traders — for in the provinces the Bank rate is the effective rate, whether it is so in London or not — the directors cannot be blamed for deciding to afford relief at the earliest opportunity."

The *Economist* says further: "The Bank rate is the rate which governs monetary business in the provinces, and when that rate is kept high above the market value of money, provincial traders are penalised." August 23, 1890, p. 1074.

In 1891 it points out that a high discount rate at the Bank was unwarranted: "To maintain it was simply to impose a needless burden upon trade, for fictitious although [sic] the 4 per cent. rate was, it was, nevertheless, the rate by which the banks all over the country regulated their charges." *Economist*, June, 1891, p. 785.

Easton notes that "When the Bank of England rate is high, the traders of this country suffer somewhat, because most of the banks are guided by the bank rate in fixing their charges." H. T. Easton, *Banks and Banking*, London, 1896, p. 146.

In 1905 the Chairman of the London and Southwestern Bank, Limited, spoke of the relations of the Bank rate to charges for advances as follows: "Now, as the larger proportion of the available funds of bankers is advanced to customers on terms varying with the Bank rate, it is obvious that in most banks, including our own, a high Bank rate increases the gross income from advances, and it is so in a greater proportion than the offset produced by a higher rate of interest on deposits, although this latter rate ordinarily changes automatically with the Bank rate." *Bankers' Magazine*, March, 1905, p. 458.

At the branches of the Bank of England, the rate charged was that current in the provinces, according to Palgrave, *op. cit.*, p. 32.

In reply to a question of the National Monetary Commission concerning discount rates at the branches, the Governor and Directors of the Bank said: "The rates current in London are telegraphed each morning to the branches for their guidance." *Interviews*, p. 22.

See also Feavearyear, *op. cit.*, p. 293, and *Macmillan Report, op. cit.*, par. 70.

[2] See Table XIV and Chart XVI. A mass of testimony with respect to the situation in the country is summed up by the statement of the Chairman of the London and Midland Bank, Limited, at the meeting in January, 1905: "In former years the rates for bankers' deposits in the provinces were regulated by the rise and fall of the Bank rate, but of recent years this had not been the case." Rates in London continued to vary with the Bank rate, however. *Bankers' Magazine*, March, 1905, p. 461.

the competition of the Post-Office Savings Banks and municipal corporations prevented the local banks from lowering their rates below $2\frac{1}{2}$ per cent.[1] Some banks held to a policy of fixed interest charges on loans,[2] and their deposit rates were also fairly steady. Even when country banks did not hold to a steady loan rate, their charges for advances varied less than those of London banks, and much less than rates for discounts.[3] A definite minimum rate was generally held to in all cases, however.

It was generally admitted that profits of provincial banks were more stable than those of London banks,[4] and this difference between City and country banks was a major factor in inducing bank amalgamations.[5] It was particularly apparent in the years

[1] Cf. the speech of the Chairman of the London City and Midland Bank, Limited, at the meeting in January, 1909. *Ibid.*, March, 1909, p. 469.

[2] Even after the low rates of the middle nineties had drawn down charges of all types this rate was not usually less than 3 per cent in London and 4 per cent in the country, ". . . and this means that when Bank rate declines below this level the rate for business conducted upon these terms does not fall further." *Ibid.*, July, 1905, p. 3.

[3] "In the great provincial cities, second only to London itself in the magnitude of their commercial and banking operations, the rate of interest will follow more or less closely the London rate, both in respect of interest on deposits and the discount on bills: but in the smaller town and country districts, a less fluctuating scale of rates is followed and preferred." George Rae, *The Country Banker — His Clients, Cares and Work*, London, 1885, p. 131. I have used here the American edition (New York, 1886).

"Your clients are accustomed to a less spasmodic scale, and would rather work upon one that seldom lost sight of 5 per cent very far either way, as a rallying point. In fact, your terms of discount must be governed, more by the expenses of your branch, than by the movements of bank rate. . . ." *Ibid.*, p. 137.

"The balance on an overdrawn account, therefore, being of the nature of a permanent loan, will not follow the ups and downs of London rate, except at a respectful distance." *Ibid.*, p. 138.

He shows that such a local bank has a very small turnover of bills, and, therefore, cannot be expected to discount at the low rates of a London bank, with many times the number of transactions. *Ibid.*, pp. 136–137.

[4] The following statements are typical: "The banks suffering a falling off in profits appear to be those which are more especially affected by city movements, while the banks having larger suburban business are able to record an increase of profit, although somewhat small." *Bankers' Magazine*, February, 1892, p. 202. "The banks whose dividends fluctuate most widely are those which confine their business chiefly to the City." The London and Westminster Bank, Limited, is cited as a good example of the purely City bank: ". . . we find consequently that its dividends are more liable to fluctuation than those of banks transacting a very large country business." *Statist*, July 2, 1904, p. 21.

[5] Sykes makes much of this difference in explaining the growth of large banks. The low money rates in London during the nineties caused the London banks to seek

following the Baring Crisis; for while the purely metropolitan banks found their earnings greatly reduced, the country banks continued to earn large profits. In this respect the banks with country connections added to a City clientele were more fortunate than banks operating almost entirely in the London market.

The London banks which kept a considerable proportion of their assets in the shape of bills found that the earnings varied with the "profit margin"— the difference between the market rate for fine bills and the deposit rate.[1] When the supplies of "money" ran short, the market rate was high, and profits large. In fact, whenever the market rate was in close accord with the Bank rate the banks gained, for the deposit rate fluctuated with the Bank rate. At the beginning of the period treated the rate on deposit balances was normally fixed at 1 per cent below the rate at the Bank, but as time passed the spread became greater, averaging $1\frac{1}{2}$ per cent, and finally reached 2 per cent for the most part in 1922.[2] The change was due largely to the ineffectiveness of the Bank rate over considerable periods. At times the profit margin was reduced to less than nothing,[3] and the banks gradually reduced their deposit allowance in consequence. Eventually they agreed not to make "floating" loans at less than the deposit rate.[4]

Country banks, on the other hand, kept their assets more largely in advances and overdrafts. Their bills, moreover, were primarily what are known as "inland" or trade bills, and these bore a higher rate than the bankers' bills discounted in London.[5]

alliances with country banks in order to have alternative fields of investment. "Since money rates in the provinces were not only more remunerative but also more stable, it was natural for the London banks to seek alliances with the banks having a preponderance of such business." There were eighteen such amalgamations in the five years following 1890. *Op. cit.*, p. 48.

The dividends of banks did not vary so widely, since it was the policy of many to maintain stable returns to shareholders. Large reserves were accumulated in years of prosperity to enable them to pay the same dividend in less prosperous times.

[1] City banks allowed interest only on deposit accounts. Consequently the gradual growth of current as compared to deposit accounts increased the earnings of City banks, and the profit margin became of less significance.

[2] See *Bankers' Magazine, passim,* and Spalding, *op. cit.*, p. 76.

[3] This was true, for example, in the first half of 1894. The average rate on deposits was *above* the bill rate. See the Chairman's speech at the meeting of the Union Bank of London, Limited, in August, 1894, *Bankers' Magazine*, September, p. 269.

[4] See above, p. 150.   [5] See Rae, *op. cit.*, p. 135.

When the demand for accommodation from banks was large they were in a position to raise their rates. This was possible on account of the local monopoly enjoyed by many of the country banks. In view of the difference between profits of provincial and City banks, it is unlikely that rates on loans in the country varied closely with the Bank rate.[1] But it is probable that the provincial banks tended to raise their charges when the Bank rate rose appreciably. Furthermore, the application of a fixed minimum prevented local charges from falling very far in periods of low Bank rates.[2] On the other hand, after 1900 the country banks suffered somewhat because deposit rates did not fall far on account of the competition of local bodies.

The country banks were able to maintain local monopolies and charge high rates in periods of depression by virtue of their relationship to London banks. The provincial banks had agents in

[1] The data on average rates given to the National Monetary Commission show much less fluctuation in country rates. See Table XV and Chart XX.

[2] There is much testimony to the effect that country rates were much higher than City rates in periods of depression. The following extracts are typical: "Those banks which have country connections, and can charge customers for accommodation, whether for loans or remittances, without much regard to the Bank rate, may escape easily, others, which depend on London business, can hardly keep up their dividends." *Bankers' Magazine*, July, 1892, p. 26.

"All banks, doubtless, have suffered more or less from the prevailing depression, but the extent appears to have been in proportion as the nature of the bulk of each bank's business is what may be termed 'City' business or otherwise, those having branches in suburban and outlying districts being able to secure loans and discount terms higher than the City market rate, while taking advantage of the minimum deposit allowance." *Ibid.*, August, 1892, p. 153.

"In the country at the present time the normal rate for three months' trade paper is 3½ to 4 per cent. For paper of similar currency in London the banks are getting ¾ per cent." Such differences were absurd, for much of the provincial paper, while not first-class, was of undoubted quality. *Ibid.*, August, 1895, p. 130.

The Chairman of the Birmingham District and Counties Banking Company, Limited, speaking at the meeting in January, 1892, remarked on the unfavorableness of the past year for banking companies. "It had affected them, however, less in the country than it had the London bankers, as they had found favourable opportunities of making temporary advances at their head office and branches at better rates than in London, and thereby extending their connection on strictly banking lines." *Ibid.*, March, 1892, p. 438.

The Chairman of the London and Provincial Bank, Limited, at the meeting in January, 1895, said it was unnecessary to point out that their business " . . . was that of a country bank, and that they supplied the financial needs of the agricultural community and small traders and private persons all over the country; therefore the causes which operated to the disadvantage of London bankers did not affect them in a like degree." *Ibid.*, February, 1895, p. 300.

the City, with whom balances for clearing and reserve purposes were kept. This system provided a channel through which surplus funds could be sent to London for investment in depression,

TABLE XV

AVERAGE RATES FOR DISCOUNTS AND LOANS IN LONDON AND OTHER ENGLISH CITIES AND TOWNS *

(Per cent)

| | London | | | Country Towns | |
| | Discounts | | Advances and Overdrafts | | |
Year	Bank of England Rate	Fine Remitted Bills and Inland Paper of the First Class 60 days	To S/E a/c to a/c	Discounts and Inland Bills ‡	Advances and Overdrafts §
1888	3.30	2.30	†	3.90	4.40
1889	3.58	2.40	†	4.00	4.50
1890	4.52	3.50	†	4.60	5.10
1891	3.28	2.44	†	3.70	4.20
1892	2.51	1.42	†	3.50	4.00
1893	3.05	2.04	†	3.60	4.10
1894	2.10	.95	†	3.50	4.00
1895	2.00	.79	†	3.50	4.00
1896	2.48	1.44	†	3.60	4.10
1897	2.63	1.80	†	3.50	4.00
1898	3.25	2.53	3.33	3.70	4.20
1899	3.74	3.19	3.89	4.00	4.50
1900	3.94	3.62	4.26	4.00	4.50
1901	3.70	3.12	4.02	3.90	4.40
1902	3.32	2.93	3.70	3.60	4.10
1903	3.77	3.18	4.14	3.70	4.20
1904	3.29	2.66	3.33	3.50	4.00
1905	3.00	2.59	3.46	3.60	4.10
1906	4.26	4.01	4.91	4.10	4.60
1907	4.90	4.44	4.91	5.00	5.50

* SOURCE: (United States) National Monetary Commission. *Statistics for Great Britain, Germany, and France, 1867–1909*, 61st Congress, 2nd Session, Senate Document No. 578, p. 143.
† The rate for these years may safely be taken as 0.25 to 0.50 per cent above bank rate.
‡ These are averages of the finest rates for the class of business. The general averages would be a little more, but would show a declining tendency in the twenty years.
§ Purely agricultural advances never less than 5 per cent. Overdrafts rather more than advances.

and withdrawn again when business expanded in the country.[1] The amplitude of the fluctuations of money rates in the London market was increased as a consequence of this movement of local

[1] Evidence to this effect is found continuously throughout the period studied in the *Economist* and the *Bankers' Magazine*. The following is typical: "The true solution of the increase in the deposits of the metropolitan joint stock banks probably is,

funds. But it was advantageous to the local banks, for they gained from the high rates imposed on local loans, and secured some profit from the funds employed in London. The situation

CHART XX

AVERAGE RATES FOR DISCOUNTS AND LOANS IN LONDON AND OTHER
ENGLISH CITIES AND TOWNS

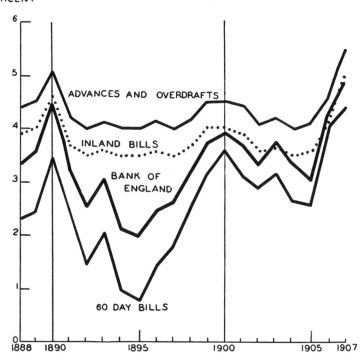

PERCENT

ADVANCES AND OVERDRAFTS

INLAND BILLS

BANK OF
ENGLAND

60 DAY BILLS

was closely analogous to that seen in the case of monopoly dumping.

that owing to this very depression of trade, and to the consequent inability of country bankers to employ their money profitably locally, they have allowed it to accumulate in the hands of their London agents." *Economist,* Commercial History and Review of 1883, p. 42. The following references are particularly apt. *Bankers' Magazine,* December, 1888, pp. 1254–1255; May, 1895, p. 707; August, 1895, p. 129; March, 1898, p. 392; July, 1911, p. 3; and January, 1913, p. 31. See also John B. Martin, "The Evolution of Our Banking System," *Economic Journal,* vol. I, no. 3 (September, 1891), pp. 543–544.

It would be natural for industrial companies in the country to discount their bills or to secure advances in the City, rather than pay higher charges in the provinces during depression. This movement certainly took place, especially when the depression was long-continued.[1] But banks were not easily induced to grant advances to new customers who were not regular borrowers, and the discounting of bills for other than their own customers was frowned upon.[2] Moreover, in depression the profit accruing to banks on bills was almost negligible, and they were disinclined to take bills from strangers, with whom the risk of loss was increased.[3]

The placement and withdrawal of funds in London by country bankers, then, was an easy process. But the other half of the mechanism, which would have kept country rates down, namely, the borrowing or discounting from London banks by local companies, was largely absent. It worked if given sufficient time, but even then only half-heartedly.

The amalgamation movement in banking brought important changes, however, for the larger customers of a branch of one of

[1] This was especially true of the latter years of the depression of 1891–1895. The quotation below is indicative:

"It is a common thing nowadays to see provincial corporations going to London banks for the money they require. A mere difference of ½ per cent. has sent Parish and County Councils to London, although the rate they refused was low enough of itself." *Bankers' Magazine*, August, 1895, p. 120.

In answer to questions of the National Monetary Commission, Sir Felix Schuster said that customers might get accommodation at branches of their bank at the Bank rate, although there might be a minimum of 3 or 3.5 per cent. He also stated that bill brokers went all over the country and discounted the paper of good traders, as they knew that the Bank of England would take it. *Interviews*, pp. 53–54.

[2] See Rae, *op. cit.*, p. 248.

[3] The Chairman of the City Bank of London, Limited, made the following statement at the meeting in July, 1895: "Their discounts were slightly less. They had met their customers' requirements, but the rate had not been tempting." *Bankers' Magazine*, August, 1895, p. 243.

Undoubtedly other banks felt the same way, for the profits to be secured on such discounts were small, and the risks involved in new customers would be greater than on their other advances or discounts.

In periods of depression the purchase of securities, or the lending of funds to stock exchange operators, was more profitable to the banks than the discount of bills. This policy is evident in the various years of easy money. The Chairman of the Union Discount Company of London, Limited, sums it up as follows: "When money was easy it was profitable to hold more securities and fewer bills." *Ibid.*, February, 1899, p. 242.

the large London banks were now in better position to demand the privileges available to London borrowers than were the same customers dealing with a purely local bank. Competition between the branches of the large banks tended to aid local industry, and the local monopolies of banks were broken down[1] to a considerable extent. This movement was particularly important in the nineties, and by 1900 banking in England was in the hands of a relatively small number of banks. During the succeeding years even larger amalgamations took place, and the number of banks was further reduced. Competition became keener throughout the country, and local monopoly rates became rare. The same movement of balances between London and the country persisted, however, but the flow was primarily an intra-bank movement.[2]

(5) The structure of rates in England during the earlier period of study lent itself to control by the Bank of England. In periods of expansion the advance of the Bank rate led in time to higher rates for charges, particularly in the country. Still more important was the manner in which banking funds were shifted from the country to London. As the need for funds in the country increased, balances were withdrawn, and rates in the London market tended to rise as a consequence. In depression, on the other hand, local rates remained high, thereby making production more difficult, and balances were shifted to the glutted money market. Eventually, the purchase of securities was stimulated,[3] and industrial expansion induced as a consequence.

To a certain extent the control of the Bank was aided during periods of prosperity by the advance of rates on advances in con-

---

[1] Sykes holds that the combination of banks brought more uniform and lower rates over the country. "While, therefore, it is improbable that loaning and interest rates ever varied so widely in the more compact banking area of England as in the heterogeneous divisions of America (where rates in the pre-federal reserve [sic] bank system varied as much as from 2–6 per cent.), undoubtedly the differences were sufficient to be economically undesirable." *Op. cit.*, p. 149.

[2] Marshall pointed out in 1899 that branch banking brought a quicker return of currency in the spring and autumn. *Op. cit.*, p. 282.

[3] Giffen presents a logical analysis of the reasons for increased security purchase in depression. See Robert Giffen, *Stock Exchange Securities: An Essay on the General Causes of Fluctuations in their Prices*, London, 1879, esp. p. 146.

formity with rises in its own rate. But this was a very slow-acting mechanism. The drain of funds from London led to high rates in the City, but this too was slow. Nevertheless, the power of the Bank rate over country transactions was of value in the earlier years, when the Bank had difficulty in controlling the London market. The evolution of large banking systems brought country rates more nearly into line with those of the City, and the Bank derived little aid from the former mechanism. But this same movement also brought more scientific banking, and more responsibility as power was concentrated in the hands of a few banks.[1] When necessary the Bank could exercise a potent control over them by bringing public opinion to bear, and thus supplement its mechanistic control over the money market.

During the latter part of the nineteenth century the gold reserves of England were very scanty. Many proposals were made to increase them,[2] but little action was taken until the years immediately preceding the Great War, when the large joint stock banks began to accumulate stores. The main reason for the lack of action was the expense involved. Each of the chief parties involved — the Bank of England, the joint stock banks, and the Government — sought to evade the burden. Consequently, for most of the period treated, England was "trading on a very thin equity" in the matter of gold reserves.[3]

But positive advantages resulted from this condition. The Bank was forced to take energetic measures to control the money market when its Reserve fell. With the exception of the years

---

[1] "A comparison of the methods of the private bankers who functioned in local spheres and issued excess local notes with no regard to the position from a national perspective, and with certainly no regard to an adequate reserve, with those of the highly centralized joint-stock companies not working in the dim light of privacy, but in the glaring light of publicity, knowing well that a deviation from policy commands instant criticism, reveals a great contrast." Sykes, *op. cit.*, p. 122.

[2] The chief proposals are summarized by Hulftegger, *op. cit.*, pp. 264–265.

[3] In 1888 Marshall suggested that the Fiduciary Issue should be increased, and that £1 notes should be issued. The Bank would be expected to keep £20,000,000 in reserves beyond the requirements of current business in ordinary times. Gold exports to an amount of £5,000,000 or £6,000,000 could then be made without necessitating violent efforts on the part of the Bank to affect the money market. *Op. cit.*, pp. 110–113.

Later on (1899) Marshall said that the Bank had earned the gratitude of the country for increasing the ordinary reserves held by them. *Op. cit.*, pp. 323–324.

from 1894 to 1897, the Bank never had large excess reserves.[1] As a result, the Bank resorted to various devices in an attempt to keep market rates in accord with the Bank rate during periods of expansion. Especially in the earlier years of the period studied, the Bank was aided to some extent by the rise of rates in the country. Gradually the growth of large banks through combination and establishment of new branches, and the development of central banking technic, enabled the Bank to regain any power over the money market which it had lost during the period of heavy gold imports into the country. To the extent that it is desirable to curb a period of prosperity early in order to avoid a long and severe depression, scanty gold reserves were advantageous. With the growth of larger foreign balances in London, however, it became increasingly desirable to have excess specie, in order that international demands for the return of balances should not upset the local credit situation.

[1] See Chart IX above, and Mitchell, *Business Cycles*, pp. 377–379.

CHAPTER IX

GOLD MOVEMENTS AND BANKING POLICY

(1) Cyclical fluctuations in the movement of gold to and from England occurred in the pre-war era. There was a tendency for imports to increase during the prosperity stages of business cycles, and for exports to grow during depression. The same phenomena were evident in the United States from 1896 to 1913.[1] Coupled with the fluctuations in foreign movements of gold were important variations in the internal movements of currency in both countries. Domestic requirements for cash increased during prosperity, and there was a large flow of cash from the money centers in each country to the interior. In depression the flow was abated greatly or reversed.

These cyclical fluctuations in gold and currency movements might be explained in terms of the classical theory of gold movements. Commodity prices and the international movement of goods must dominate the situation if this explanation is to be adequate. Hawtrey's analysis of specie flows over cyclical periods is based upon the assumption that prices of goods will be the dominating factor. Gold flows because the price levels of one country do not rise or fall in accordance with the levels in other countries. The movement of gold forces all countries to keep the same pace through the various phases of the business cycle. If the price levels in England and the United States had been less sensitive than prices in other countries, the cyclical fluctuations in gold flows found here could be explained along the lines of the classical theory. In prosperity their price levels would have risen more slowly than prices elsewhere in the world, and specie would have flowed into the two countries. In depression the price levels would have fallen more slowly also, and gold exports to countries whose prices were falling faster would have taken place.

---

[1] See Appendix A below.

Undoubtedly there is some tendency for the price and wage structures of advanced industrial countries, such as England during the period studied, to be more rigid than the structure in countries which are newer in their stage of economic development. On the other hand, there is perhaps a tendency for the elasticity of demand for commodities produced for export by "older" nations to be somewhat greater on the average than the elasticity of demand for products imported by "old" countries from "newer" nations. If prices of all commodities entering into foreign trade were to rise by the same degree during the prosperity phase of a world business cycle, the total sums due by "older" nations to "young" countries would rise *relatively* to the amounts owing to them by the latter. Gold would then flow to the "young" countries instead of to the more developed nations. A considerable difference in the rate of price change in the two sets of countries would be necessary in order to provoke flows in the opposite direction. In depression the sums due to the "young" nations would decline greatly, and the amounts owed to "older" countries would rise *relatively*. Gold movements to the more developed nations would take place. If the fall of prices in the "newer" countries was relatively great, the specie movement would be even larger.

Shifts in demand for goods in a schedule sense, however, are likely to be more important than the elasticity of demand, when price and income levels are changing. Shifts in demand schedules occur not only because of price changes in general, but also because of fluctuations in the volume of foreign loans. The flotation of loans enables the borrowing nations to increase purchases of goods without suffering a loss of gold. The volume of new loans tended to increase during prosperity and to fall off in depression in the case of England.[1] There would be a tendency for gold to be lost, as a consequence, unless goods movements expanded and contracted in proportion to the loans. If the price level of England did not rise as rapidly as in the rest of the world, goods exports would increase during the periods of prosperity. But this would merely balance the increase in foreign loan flotations in

[1] See below, pp. 175-176.

part, and gold would flow into the country because of the goods exports induced by the price differentials between England and the rest of the world. Any excess of the volume of foreign loans above the expansion in exports of commodities would tend to bring specie exports.

In less developed countries commodity imports would increase with borrowing operations. Some specie imports would occur if the proceeds of the loans were not taken wholly in the form of goods. The expenditures from the loans would tend to increase the price level in these countries relatively to levels in creditor nations. But no specie would be lost, so long as the borrowing operations continued.

There are two principal difficulties with the classical analysis in connection with cyclical gold flows. In the first place, the adjustment through the price mechanism must be very prompt. International trade must respond quickly to changes in the general level of prices. Over periods so short as the business cycle it is not certain that these adjustments will be made rapidly enough for gold movements to play the role assigned to them. Secondly, the fluctuations in foreign loans should permit relatively higher price levels in borrowing countries during the prosperity phases of the cycle, and should bring a greater fall in these levels during depression, as compared with creditor nations, if the classical explanation of adjustment in the case of foreign loans is correct. Even though the price levels were lower in lending nations than in borrowing countries, gold would flow to the latter. Again, however, the adjustment of goods movements to loan operations takes time, and it may not be perfect over so short a period as the business cycle. The chief defect with the classical explanation is that lower price levels in prosperity in the "older" nations, which are also the creditor nations, would lead to specie imports if price levels govern the situation, but would be consistent with specie exports if there was a large volume of international lending occurring. A lag of prices in these countries during the expansion and prosperity phase of the cycle should not necessarily bring specie imports. In depression a lag of prices would lead to gold exports, but this movement would be counteracted by a decline in the volume of foreign loans. Net specie exports could occur only if the price level fell much faster than in other nations. Dependence cannot then be placed upon an insensitive price structure as the explanation for the cyclical gold movements found for England.

It is obvious that foreign loans would affect debtor and creditor nations in different ways, so far as gold flows are concerned. But the cyclical movements were similar for both England and the United States in the pre-war period. The United States might have had a more sensitive price structure than England, and in that case gold exports would have occurred in prosperity, and imports would have taken place in depression. In prosperity the borrowing operations would have tended to bring in gold to offset the specie exports due to relatively high prices, while in depression the decline in loans would have prevented specie imports. It is difficult to make out an *a priori* case for specie movements over cyclical periods where two important forces of this type work at cross purposes. The same situation prevails for England, since the making of loans in prosperity would have led to specie exports, but the insensitive price structure would have led to imports.

(2) In one respect, however, both nations were alike in this period. The banking systems of both countries had little slack, and relatively small changes in the volume of credit tended to reduce reserves markedly. Even more important was the fact that both used standard money coin for a large proportion of the cash transactions of the public. With rising prices in prosperity the requirements of the public increased, and this drain was more important than the increase of bank credit for the depletion of reserves. The pressure upon bank reserves led to increased discount rates in the money markets, and the flow of short-term capital, induced as a consequence, may well have been the dominating factor determining the gold flows over cyclical periods. So long as rates remained above the world levels, the funds would remain in the two centers. In time the high discount rates would affect business conditions, and the rise of prices and production curtailed. But this period might be postponed for several years. In

depression the return of gold from the interior would ease the money markets and bring about an outflow of balances accompanied by gold.

Hawtrey maintains that the response of prices and production to the rise of discount rates will be quick, because of the activities of traders operating on little capital of their own, and dependent principally upon bank credit. An increase in the cost of this credit would lead them to curtail orders. Some variation in the spread between buying and selling prices of goods might enable the traders to bear the added interest cost, however. And the traders must occupy a prominent place in the economic system if the mechanism is to work promptly. International conditions would be affected by traders in London, and fluctuations in the volume of their orders would tend to keep all nations at about the same stage in the business cycle, according to Hawtrey. If this were true, however, cyclical fluctuations in gold flows should not be very marked, since adjustment of disequilibria in the balance of payments would occur quickly through goods movements. The evidence found in this study points to very important cyclical variations in specie movements, and it seems probable that the frictional elements in the economic system are more important than the classical writers have admitted.

(3) If the internal currency requirements were the dominant factor in connection with "tight" banking systems causing the cyclical fluctuations in gold movements in England and the United States, it is clear that in addition to the cost of coinage and the loss of interest on the specie, standard money currencies are expensive because of their effects in upsetting the economic systems of other countries. Unless the requirements of the public for additional coin in periods of prosperity are met by withdrawals from hoards somewhere in the world, other nations will be forced to contract more rapidly than would have been the case. Even if the needs are supplied wholly from newly mined gold, other countries suffer, since they are unable to secure their usual share of the annual output of metal. In depression the flow of gold to other countries tends to upset their money markets, and to

produce an easy condition which is deceptive. As in the case where balances are held in foreign centers by countries on the gold exchange standard, these nations must be ready to part with their specie when times improve.

Coin circulations consisting of standard money have been advocated on the ground that they constitute "secondary" reserves.[1] Gold coin was not available for export, however, during periods of business expansion when it was most likely to be needed, and the internal drain was likely to be heaviest at times of crises. Under long-run conditions a coin circulation may constitute a secondary reserve, because it may be assumed that commodity prices will dominate the situation. If a nation should begin to lag behind other countries in its economic development, adjustment to the altering situation would occur gradually. The fall of domestic prices brought about by gold exports would release gold from the coin circulation to supplement bank reserves. But over periods so short as the business cycle, there is no *a priori* reason to expect that the necessary adjustments will be made in this frictionless manner.

(4) Before examining further the possibility that fluctuations in discount rates and short-term capital movements are the chief explanation for the cyclical movements of gold found for England and the United States, the variations in the foreign loans made by England during the period studied will be considered. In Table XVI, capital exports, earnings from shipping, and receipts from commissions for financial services are shown.[2] These data are plotted in Chart XXI, together with the Thomas index of business conditions. Capital exports declined during the period from 1881 to 1883, although business conditions were improving, and there was very little expansion in foreign loans in the years from 1895 to 1900.[3] In the other periods of business prosperity, 1886–

[1] Lord Swaytheling maintained that Viscount Goschen was responsible for originating this notion, and he denied that the coin circulation was a secondary reserve because the gold was not available when it was needed. See *Interviews*, pp. 93–94.

[2] These data are taken from C. K. Hobson, *The Export of Capital*, London, 1914, pp. 197 and 204.

[3] It is difficult to remove the secular trend from the series of capital exports. The best fit was secured by using three straight line trends for the periods 1881–1890,

1890, 1904–1907, and 1909–1911, exports of capital increased greatly. In the periods of declining business from 1890–1895, 1900–1904, and 1907–1908, capital exports fell off. The timing of the movements is not parallel in all cases, but there is a high posi-

TABLE XVI

Selected Items from the International Balance of Payments, Great Britain, 1881–1911 *

(In millions of pounds)

Year	Capital Exports	Shipping Earnings, etc.	Commission Insurance and Banking Charges	Year	Capital Exports	Shipping Earnings, etc.	Commission Insurance and Banking Charges
1881	33.2	49.0	14.8	1897	27.1	57.7	14.0
1882	24.3	51.5	16.8	1898	17.2	65.5	15.8
1883	16.9	54.0	15.8	1899	27.9	64.5	18.0
1884	41.0	50.0	13.6	1900	31.2	76.3	20.2
1885	33.9	46.0	12.6	1901	13.9	60.0	19.1
1886	61.8	45.5	12.4	1902	11.2	60.0	18.3
1887	66.8	46.0	13.5	1903	23.0	64.5	20.0
1888	74.5	52.5	14.3	1904	27.2	66.0	19.2
1889	68.8	62.0	16.0	1905	62.8	72.5	19.4
1890	82.6	57.0	18.3	1906	104.4	85.0	25.3
1891	48.5	56.5	15.5	1907	140.2	90.0	29.9
1892	35.3	52.3	13.2	1908	129.9	79.0	20.9
1893	40.1	52.0	13.7	1909	110.1	83.0	21.6
1894	21.3	52.0	11.8	1910	150.8	95.0	26.7
1895	22.7	50.2	11.9	1911	192.2	110.0	26.3
1896	39.3	54.0	13.6				

* Source: C. K. Hobson, *The Export of Capital*, London, 1914, pp. 197 and 204.

tive correlation between the condition of business and the volume of capital exports.[1]

1891–1903, and 1904–1911, respectively. The periods were too short to insure proper handling of the years at the ends of each period. Somewhat higher correlation between capital exports and business conditions was indicated by a comparison of deviations from the trends and the cyclical fluctuations shown by the Thomas index than is shown in Chart XXI, where the original data of capital exports are compared with the index of business.

[1] White finds that fluctuations in business conditions and variations in the export of capital were not positively correlated in all years. His method of measurement involved giving weights ranging from + 3 to − 3 to the classifications made by Thorp in *Business Annals* for business conditions in each year, and comparing this rough index with exports of capital. In twenty-three of the years from 1880 to 1912, the fluctuations of business and capital exports were in the same direction, while in ten years there was a negative correlation. See White, *op. cit.*, pp. 217–221.

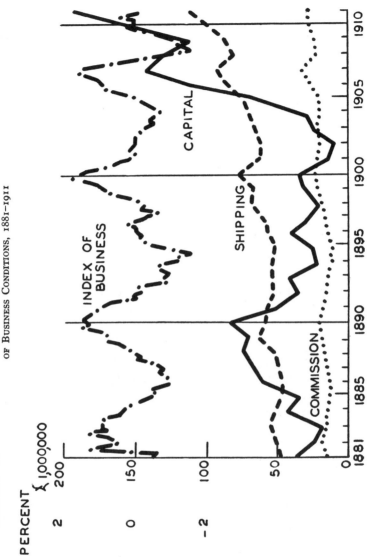

CHART XXI

ITEMS FROM THE INTERNATIONAL BALANCE OF PAYMENTS OF GREAT BRITAIN AND THE THOMAS INDEX OF BUSINESS CONDITIONS, 1881-1911

Cyclical fluctuations in earnings from shipping and commissions are evident, although the amplitude of the fluctuations is less than in the case of capital exports. Business prosperity brought increased earnings from these sources, and depression caused them to decline. The most important divergence in the curves shown in Chart XXI is in the earnings from shipping following 1901. Receipts began to increase three years prior to the improvement in business conditions.

(5) It has been pointed out by Angell that an intimate connection exists between the movements of capital and the balance of commodity trade.[1] But he denies that fluctuations in these flows can be explained in terms of the business cycle. He argues that if the cycles in the lending and borrowing nations are coincident and of roughly equal magnitude, their influence will not materially affect the situation. If the cycles are not similar, the prices of securities will fluctuate in different ways in the two countries, and a flow of capital might occur. But the prices of stocks and of fixed-yield securities will tend to move in opposite directions, and consequently there will be a cross-current of investment flows between the two nations. The latter type of security passes chiefly in international dealings, so far as new loans are concerned, however, and to the extent that fixed-yield securities are more important than stocks in international transactions, some cyclical fluctuation in investment would occur. Angell goes on to deny that the cyclical hypothesis can account for variation in other items in the international accounts, such as shipping charges, immigrants remittances, tourist expenditures, etc.

The two countries considered by Angell are presumed to be financially independent, and capital can move freely either way. In the case of the majority of long-term loans, however, this assumption cannot be made. New loans are made by nations with large volumes of accumulated capital, while the borrowers are principally newly developed countries. In a period when discount rates are rising in debtor nations, new loans will be made by creditor countries because of the higher rates, and securities of

---

[1] *The Theory of International Prices*, pp. 527–528.

the debtor nation which are already on the market will be more attractive to investors in the creditor country because of the fall in their prices. But direct investment in the debtor country is also profitable for foreign investors because of the business expansion occurring there. If the expansion is accompanied by a "boom" stock market, foreign speculators will be tempted to transfer funds to the country. Even if the creditor country is experiencing depression, the low price of stocks will not tempt investors or speculators in the debtor nation to invest, since opportunities for investment or speculation at home are at least equally tempting. But little flow of funds from the debtor country to the creditor nation would occur whatever the relation of the business cycles in the two countries to each other, because the volume of funds at the disposal of investors in the debtor country is relatively small. While some cross-currents of investment flows may well occur between financially independent nations, the vast bulk of the international movement of capital is accounted for by transactions between creditor and debtor countries. Cyclical fluctuation in this movement is probable, both because the volume of new loans is determined to a large extent by the conditions of business in the lending nations, and because psychological forces are of great importance. Expansion of loans in prosperity and contraction during depression is to be expected, as a consequence.

(6) Movements of short-term balances between financially independent nations, however, may be determined by relative differences in discount rates, rather than by the underlying conditions of business in the two countries. Unless the influence of discount rates upon commodity prices and the volume of production is felt immediately, there is no reason why such short-term loans should flow in the same direction as long-term loans. So long as the flow of short-term funds continues, the nation receiving the balances would not lose gold, where the movement was large, and it would be enabled to continue its business expansion until the high discount rates proved to be a deterrent to business, or, more important, the bankers became unwilling to lend.

The international movement of short-term balances is likely to be very sensitive to changes in discount rates, but domestic business is unlikely to respond so quickly to rate changes. Central banks are placed in a position where ordinary discount policy may be ineffective. In periods of prosperity the rise of rates designed to curb business expansion may merely bring in foreign balances, and enable the country to continue its expansion to a greater extent. And in depression the fall of rates leads to an outflow of foreign funds, and money market conditions are not eased in the manner expected.[1] Unless the general price structure is very sensitive to rate changes, which is only likely to be the case when a large proportion of the bank credit made available to business is provided by the central money market, movements of foreign balances may dominate the situation over periods so short as the various phases of the business cycle. Cylical fluctuations in the movement of specie might easily be controlled by the movements of these balances. This explanation for the cyclical movements of gold found for England and the United States seems more adequate than the explanation based upon price level differences. The existence of banking systems with little slack made the flow of foreign balances very important, because the drawing down of reserves in prosperity occurred early. Internal currency movements added to the tightness of the money markets, and rising rates tended to draw gold to the countries. During depression the reflux of funds and cash from business firms and the public produced an ease in the money markets which caused the withdrawal of foreign balances.

(7) If the nations with "tight" banking systems are not to upset the world economy, they must hold large excess reserves of gold, so that specie will not have to be drawn from the rest of the world during the stages of business prosperity. But this alone is not sufficient, since any attempt by the central banks to curb domestic business may lead to an influx of foreign balances accompanied by specie imports. In the absence of sensitive price structures within the countries, central banks may have to resort to qualitative measures of control, or rationing of credit. It is

---

[1] Cf. Keynes, *op. cit.*, Ch. 36, and Williams, *op. cit.*, esp. pp. 142–148.

possible that a scheme of rate control might be devised where rates to banks on rediscounts outside the central money market might be raised prior to the increase of rates at the center. In the United States, rates at the Federal Reserve Banks in the interior of the country would be raised before the New York Bank acted, thus reversing present-day policy. While politically difficult to apply, such a policy would have the effect of applying pressure first upon business in the interior which is unable to borrow in the major money market. Easing of rates there in depression might be quickened, and to some extent the evil effects of ordinary bank rate stability over cyclical periods would be abated. The lag of rates in the central money market behind the rates in the interior would prevent foreign balances from exerting such a powerful effect.[1] Such a mechanism assumes that the fluidity of credit is less than is generally thought. The evidence of rates and movements of banking funds in England in the period studied suggests that a considerable proportion of business firms were not able to tap the central money market, and were forced to accept the conditions laid down by local banks. Dr. Riefler's study of rates in the United States indicates that the same type of condition prevails in this country.[2] Perhaps writers on banking have been influenced too much by the fluidity of funds in money markets, and have assumed tacitly that such fluidity is also true of the banking structure in general when such has not been the case.

Co-operation of central banks may remedy the situation in part, by keeping discount rates in the various money markets more nearly in line. Unless banking systems are very much alike, and business conditions similar throughout the world, however, central banks may not be able to make rate changes effective in the money markets. And these banks may well hesitate to apply a policy which is desirable from the standpoint of the world economy, but which may be disadvantageous from a national

[1] Central bank rate changes may have little effect upon rates charged to customers, and in this event the only virtue of the policy suggested lies in its effect in delaying movement of foreign balances into or out of the central money market.

[2] Winfield W. Riefler, *Money Rates and Money Markets in the United States*, New York, 1930, Ch. IV.

viewpoint. Moreover, the transfer of funds for speculative use elsewhere may not be curbed by small changes in money rates. The flow of balances to New York in 1928 and 1929 for investment in brokers' loans or in securities would have been difficult to control even though central banks in Europe had raised rates to very high levels.

Keynes' suggestion that an element of doubt as to the future terms of exchange between currencies should be established, by providing for larger variations in the buying and selling prices of gold by central banks, and the establishment of forward exchange market rates under the control of the central banks,[1] appears to have great merit. In a world where internal price structures are rigid, the easiest method of curbing the unfortunate effects of movements of short-term balances is to set up conditions which make the latter more insensitive to rate changes. Unless such a mechanism is devised, the international gold standard seems likely to prove effective only over short periods, and to be subject to breakdown at frequent intervals.

[1] *Op. cit.*, vol. II, esp. pp. 315–331.

APPENDICES

APPENDIX A

GOLD MOVEMENTS AND BUSINESS CYCLES IN THE UNITED STATES, 1896–1913

Attention has been called earlier to the high inverse correlation found between foreign movements of gold at New York and the internal flow of currency over cyclical periods, and this phenomenon will be examined briefly here. There is abundant material for the study of business and banking conditions and the movement of specie in the United States, but most of the data have been subjected to analysis in other studies, and it is necessary only to bring together the important results for study in relation to the problems considered in the case of England.

The banking system of this country was admittedly a poor one in many respects prior to the adoption of the Federal Reserve System, and its defects were most clearly apparent in times of commercial crises. Any severe strain on the reserves of the important clearing banks in New York City was likely to bring a suspension of payments and general financial panic. The most important reasons lay in the inelasticity of reserves and the lack of a central banking institution. The interlocking of reserves, through the deposit of portions of the legal reserves of banks in outlying cities and towns in the banks in reserve and primarily central reserve cities, coupled with the fact that gold was the only elastic element in the ordinary hand-to-hand circulation, brought a heavy strain upon the cash of the country. Without a central bank to curtail credit operations during the latter stages of business expansion and to advance funds to meet crisis needs, the system operated in uncontrolled freedom during the expansion phase, only to break down completely under the demands of the crisis stress.

It is true that the United States used other forms of money than gold, but they were inelastic, and could not be expanded quickly in order to meet the needs of the country for circulating media.[1] And this was an ever present need, for the autumnal drain of cash necessary to move the crops was very large. The seasonal movements of credit and cash in response largely to agricultural requirements have been studied by Professor Kemmerer for the National Monetary Commis-

---

[1] See W. C. Mitchell, *Business Cycles*, pp. 288–292, and p. 295. Professor Allyn A. Young has pointed out that the National Bank Notes were not "perversely inelastic," as has often been maintained, but it is apparent from his study that the fluctuations in their volume are not important quantitatively. Young, *op. cit.*, pp. 25–26. Cf. Mitchell, *op. cit.*, p. 324.

sion.[1] It was recognized that crises were most likely to occur in the fall of the year, since the banks in New York were then called upon to meet the heavy internal demand at the time when they were weak. The additional strain was sufficient to bring their downfall.

In addition to seasonal demands, there was a drain of currency into circulation during the expansion phase of the business cycle. Studies of the relationship between the amount of money in the hands of banks and in the pockets of the people have been made, which show this clearly.[2] Banking reserves were drained in prosperity and replenished in depression. The recognized value of the ratio of loans to deposits of the New York clearing banks as a barometer of business conditions is silent proof of knowledge of this fact. When the proportion approached close to 100 per cent it indicated that a crisis was imminent, for it meant that deposits were being withdrawn in cash, while loans were still increasing.[3] In periods of crisis gold was drawn from abroad by the high interest rates in New York.

In the introductory chapter attention was called to the observation of Professor Young with respect to the flow of gold through New York as an entrepot. His conclusions were deduced from an examination of a chart and accompanying table in A. P. Andrew's studies for the National Monetary Commission.[4] The table showed weekly net foreign movements of gold at New York, and net movements of currency between New York and the interior of the country, as reported by the *Commercial and Financial Chronicle*. This series has been extended in Table XVII to cover the full period from 1896 to 1913, and quarterly totals computed. In addition a four-quarter moving average has been calculated to remove some of the seasonal variation, as in the case of the similar English statistics.[5] The two series have been plotted in Charts XXII and XXIII.[6] Another series has been constructed

[1] E. W. Kemmerer, *Seasonal Variations in the Relative Demand for Money and Capital in the United States*, (United States) National Monetary Commission, 61st Congress, 2nd Session, Senate Document No. 588, esp. Chs. IV and XXVI. See also Leonard L. Watkins, *Banker's Balances*, Chicago, 1929, esp. pp. 72–73 and 217–218.

[2] See O. M. W. Sprague, "The Distribution of Money between the Banks and the People since 1893," *Quarterly Journal of Economics*, vol. XVIII, no. 4 (August, 1904), pp. 513–528; Mitchell, *op. cit.*, pp. 295–300; and Young, *op. cit.*, pp. 23–25 and 64–67.

[3] Mitchell, *op. cit.*, pp. 330–332, and 345–350; Young, *op. cit.*, pp. 29–30; and W. M. Persons, "Cyclical Fluctuations of the Ratio of Bank Loans to Deposits, 1867–1914," *Review of Economic Statistics*, vol. VI, no. 4 (October, 1924), pp. 260–283.

[4] A. P. Andrew, *Financial Diagrams*, (United States) National Monetary Commission, 61st Congress, 2nd Session, Senate Document No. 509, Pl. 18; and *Statistics for the United States, 1867–1909*, (United States) National Monetary Commission, 61st Congress, 2nd Session, Senate Document No. 570, pp. 172–176 and 229–231.

[5] Nearly all of the gold movements passed through New York. There was some import of Australian gold at San Francisco, but the amounts were relatively small.

[6] A plus indicates movements to New York.

TABLE XVII

QUARTERLY NET FOREIGN MOVEMENTS OF GOLD AND INTERIOR MOVEMENTS
OF GOLD AND CURRENCY AT NEW YORK, 1896–1913 *

(In thousands of dollars)

Quarter		Foreign †	Interior †	Quarter		Foreign †	Interior †
1896	1st.....	+ 3,194	+ 46,769	1905	1st.....	−27,972	+ 54,778
	2nd....	−26,921	+ 44,129		2nd....	− 3,003	+ 33,973
	3rd.....	+31,542	− 23,856		3rd.....	+ 1,149	− 9,597
	4th.....	+22,281	+ 5,870		4th.....	+ 7,372	− 6,816
1897	1st.....	+ 27	+ 27,990	1906	1st.....	− 1,681	+ 39,380
	2nd....	−22,506	+ 43,178		2nd....	+40,963	− 49,236
	3rd.....	− 2,025	− 12,605		3rd.....	+30,850	− 8,761
	4th.....	+ 8,327	+ 21,625		4th.....	+18,144	− 19,336
1898	1st.....	+27,215	+ 17,353	1907	1st.....	+ 1,287	+ 21,050
	2nd....	+37,074	+ 9,408		2nd....	−21,839	+ 15,312
	3rd.....	+13,219	+ 3,018		3rd.....	− 5,949	+ 9,054
	4th.....	+10,107	+ 17,361		4th.....	+94,241	−132,213
1899	1st.....	+ 2,023	+ 24,404	1908	1st.....	+ 7,876	+100,170
	2nd....	−19,190	+ 26,724		2nd....	−43,154	+106,967
	3rd.....	+ 1,178	− 6,140		3rd.....	+ 1,307	+ 32,991
	4th.....	− 7,286	− 8,716		4th.....	− 4,385	+ 33,782
1900	1st.....	− 6,408	+ 34,845	1909	1st.....	−33,877	+ 82,787
	2nd....	−19,636	+ 35,545		2nd....	−20,972	+ 77,343
	3rd.....	−19,500	+ 1,689		3rd.....	−15,078	+ 34,202
	4th.....	+ 8,440	− 19,416		4th.....	−16,284	+ 24,640
1901	1st.....	− 8,057	+ 53,836	1910	1st.....	− 3,525	+ 74,694
	2nd....	−18,454	+ 27,992		2nd....	−35,429	+ 77,922
	3rd.....	+ 182	− 4,659		3rd.....	+13,868	+ 43,226
	4th.....	−20,196	+ 404		4th.....	+ 2,217	+ 15,605
1902	1st.....	−13,349	+ 24,350	1911	1st.....	+ 1,174	+ 79,738
	2nd....	− 2,309	+ 29,794		2nd....	+ 3,620	+ 93,738
	3rd.....	− 7,525	+ 412		3rd.....	+ 2,175	+ 62,245
	4th.....	+ 577	− 1,610		4th.....	− 1,817	+ 67,861
1903	1st.....	− 208	+ 35,852	1912	1st.....	−13,951	+118,233
	2nd....	−21,668	+ 49,840		2nd....	− 1,750	+107,526
	3rd.....	− 6,927	+ 23,641		3rd.....	+ 507	+ 60,382
	4th.....	+18,112	− 22,403		4th.....	+15,054	+ 39,011
1904	1st.....	+ 1,651	+ 48,252	1913	1st.....	−42,310	+132,469
	2nd....	−59,706	+ 69,771		2nd....	− 8,377	+122,335
	3rd.....	− 8,746	− 471		3rd.....	− 1,682	+ 97,164
	4th.....	−29,372	+ 15,033		4th.....	+ 8,983	+ 83,560

* SOURCE: Derived from weekly reports of the *Commercial and Financial Chronicle*, New York.
† + = to New York; − = from New York.

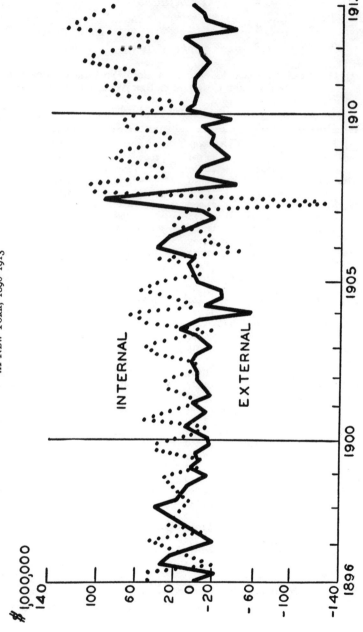

CHART XXII

QUARTERLY NET FOREIGN MOVEMENTS OF GOLD AND INTERIOR MOVEMENTS OF GOLD AND CURRENCY
AT NEW YORK, 1896–1913

CHART XXIII

FOUR-QUARTER MOVING AVERAGES OF NET FOREIGN MOVEMENTS OF GOLD AND INTERIOR MOVEMENTS OF GOLD AND CURRENCY AT NEW YORK, 1896–1913

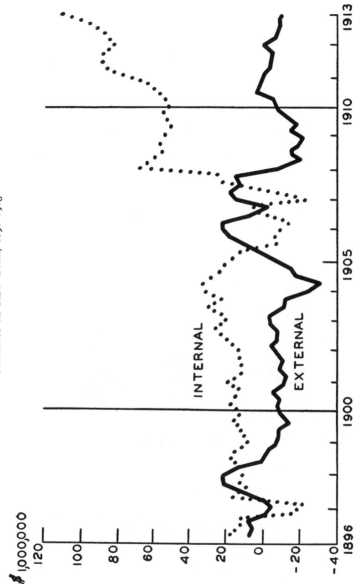

TABLE XVIII

FOUR-QUARTER MOVING AVERAGES OF NET FOREIGN MOVEMENTS OF GOLD
AND INTERIOR MOVEMENTS OF GOLD AND CURRENCY AT
NEW YORK, 1896–1913 *

(In thousands of dollars)

Centered at	Foreign †	Interior †	Centered at	Foreign †	Interior †
1896 July 1..	+ 7,524	+ 18,228	1905 Jan. 1..	−17,273	+ 25,828
Oct. 1..	+ 6,732	+ 13,533	April 1..	−14,799	+ 23,547
1897 Jan. 1..	+ 7,836	+ 13,296	July 1..	− 5,614	+ 18,084
April 1..	− 556	− 16,108	Oct. 1..	+ 959	+ 14,235
July 1..	− 4,044	− 20,047	1906 Jan. 1..	+11,951	− 6,567
Oct. 1..	+ 2,753	+ 17,388	April 1..	+19,376	− 6,358
1898 Jan. 1..	+17,648	+ 8,945	July 1..	+22,069	− 9,488
April 1..	+21,459	+ 12,851	Oct. 1..	+22,811	− 14,071
July 1..	+21,904	+ 11,785	1907 Jan. 1..	+ 7,110	+ 2,066
Oct. 1..	+15,606	+ 13,548	April 1..	− 2,089	+ 6,520
1899 Jan. 1..	+ 1,539	+ 17,877	July 1..	+16,938	− 21,699
April 1..	− 1,470	+ 15,587	Oct. 1..	+18,582	− 1,919
July 1..	− 5,819	+ 9,068	1908 Jan. 1..	+13,254	+ 20,995
Oct. 1..	− 7,927	+ 11,678	April 1..	+15,068	+ 26,979
1900 Jan. 1..	− 8,038	+ 13,884	July 1..	− 9,589	+ 68,478
April 1..	−13,207	+ 15,841	Oct. 1..	−20,027	+ 64,132
July 1..	− 9,276	+ 13,166	1909 Jan. 1..	−14,482	+ 56,726
Oct. 1..	− 7,188	+ 17,914	April 1..	−18,578	+ 57,026
1901 Jan. 1..	− 9,393	+ 16,025	July 1..	−21,553	+ 54,743
April 1..	− 4,472	+ 14,025	Oct. 1..	−13,965	+ 52,720
July 1..	−11,631	+ 19,393	1910 Jan. 1..	−17,579	+ 52,864
Oct. 1..	−12,954	+ 12,022	April 1..	−10,342	+ 55,121
1902 Jan. 1..	− 8,918	+ 12,472	July 1..	− 5,717	+ 52,862
April 1..	−10,845	+ 13,740	Oct. 1..	− 4,543	+ 54,123
July 1..	− 5,651	+ 13,236	1911 Jan. 1..	+ 5,220	+ 58,071
Oct. 1..	− 2,366	+ 16,112	April 1..	+ 2,296	+ 62,828
1903 Jan. 1..	− 7,206	+ 21,124	July 1..	+ 1,288	+ 75,889
April 1..	− 7,057	+ 26,931	Oct. 1..	− 2,493	+ 85,513
July 1..	− 2,673	+ 21,732	1912 Jan. 1..	− 3,836	+ 88,966
Oct. 1..	− 2,208	+ 24,833	April 1..	− 4,253	+ 88,501
1904 Jan. 1..	−11,718	+ 29,815	July 1..	− 35	+ 81,288
April 1..	−12,172	+ 23,787	Oct. 1..	− 7,125	+ 84,847
July 1..	−24,043	+ 33,171	1913 Jan. 1..	− 8,782	+ 88,549
Oct. 1..	−31,449	+ 34,778	April 1..	− 9,329	+ 97,745
			July 1..	−10,847	+108,882

* SOURCE: Table XVII.
† + = to New York; − = from New York.

from data appearing in Professor Mitchell's *Business Cycles* (1913). This series, which is shown in Table XIX and Chart XXIV, makes allowance for gold used in industry, and for the production of new metal, but the figures are based on annual data, and do not indicate the timing of movements as well as the quarterly series. The division of

CHART XXIV

CHANGES IN MONETARY GOLD STOCK IN THE UNITED STATES, 1890-1910

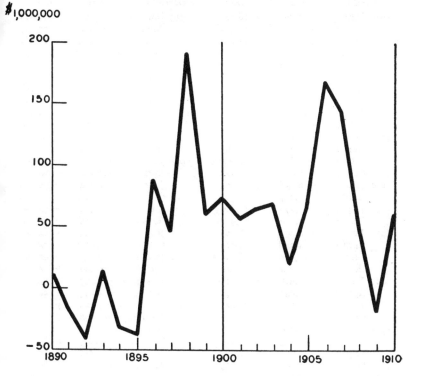

the total monetary stock among the Treasury, the banks, and the public is shown in Table XX and Chart XXV.

Foreign movements of gold were somewhat less important for the United States, because of the large domestic production. Domestic demands were supplied primarily from the newly mined supplies, and ordinarily some was exported as well. Only occasionally was it necessary to secure gold from abroad in large amounts. There was an important seasonal flow of specie, but there was also a definite cyclical

variation in the movement of gold at New York. From 1896 to 1898
there was a very heavy net import into the country, due primarily to
the large agricultural exports in these years.[1] No important foreign
flows aside from the usual export of newly mined gold occurred there-

TABLE XIX

MONETARY GOLD STOCK AND MOVEMENTS OF MONEY INTO AND OUT OF
CIRCULATION IN THE UNITED STATES, 1890–1911 *

(Millions of dollars)

| Year | Actual Amounts | Increase or Decrease | Chief Sources of Gain or Loss | | | Net |
			U. S. Production	Net Foreign Flow	Industrial Consumption	
1890	648	+ 15	33	− 4	15	+ 14
1891	626	− 22	33	− 34	15	− 16
1892	582	− 44	33	− 59	15	− 41
1893	591	+ 9	36	− 7	13	+ 16
1894	539	− 52	40	− 81	10	− 51
1895	503	− 36	47	− 71	12	− 36
1896	589	+ 86	53	+ 46	11	+ 88
1897	638	+ 49	57	..	11	+ 46
1898	832	+194	64	+142	11	+193
1899	897	+ 65	71	+ 6	13	+ 60
1900	989	+ 92	79	+ 13	17	+ 73
1901	1,050	+ 61	79	− 3	19	+ 56
1902	1,121	+ 71	80	+ 8	20	+ 65
1903	1,192	+ 71	74	+ 21	23	+ 71
1904	1,217	+ 25	80	− 36	24	+ 21
1905	1,288	+ 71	88	+ 3	23	+ 63
1906	1,458	+170	94	+109	28	+170
1907	1,605	+147	90	+ 88	33	+144
1908	1,654	+ 49	95	− 31	34	+ 49
1909	1,638	− 16	100	− 89	15	− 19
1910	1,709	+ 71	96	..	30	+ 62
1911	1,797	+ 88	..	..	34	..

* SOURCE: W. C. Mitchell, *Business Cycles, Memoirs of the University of California*, vol. 3, Berkeley, California, 1913, pp. 283–285.

after until the expansion period preceding the crisis of 1907. In 1904
there was an unusually large efflux, but the crises of 1901 and 1903
seem to have affected gold flows very slightly on the whole. Undoubt-

[1] See A. P. Andrew, "The Influence of the Crops upon Business in America," *Quarterly Journal of Economics*, vol. XX, no. 3 (May, 1906), p. 347, and C. J. Bullock, J. H. Williams, and R. S. Tucker, "The Balance of Trade of the United States," *Review of Economic Statistics*, July, 1919, reprinted in F. W. Taussig, *Selected Readings in International Trade and Tariff Problems*, Boston, 1921, p. 189.

edly, the large imports of the late nineties enabled the country to pass through these crises without drawing upon foreign countries to any great extent.[1] In 1906 and 1907 the import of gold was very large,

CHART XXV

PROPORTIONS OF MONETARY STOCKS HELD BY THE TREASURY, BANKS, AND PUBLIC, 1890–1911

PERCENT

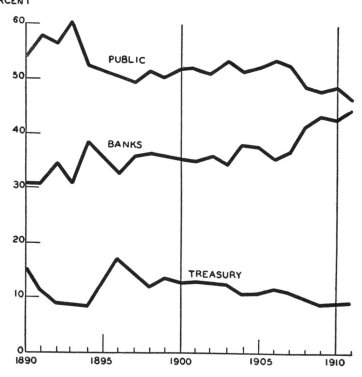

while in 1908 the movement was reversed, and the specie secured during the crisis was returned to Europe.

Internal movements show a normal flow of currency into New York during depression periods, and a drain into circulation in prosperity.[2]

[1] See O. M. W. Sprague, *History of Crises under the National Banking System,* (United States) National Monetary Commission, 61st Congress, 2nd Session, Senate Document No. 538, pp. 217–218.

[2] The heavy movements into New York from 1908 to 1912 may be explained in part by a shift from inter-bank deposits to direct lending by country banks in the New York market. See Watkins, *op. cit.,* pp. 74–76.

When allowance is made for the utilization of domestically produced metal, the drain preceding the crisis of 1903 becomes evident; and in the years immediately preceding the crisis of 1907 it is marked. It is clear from an examination of the chart showing the division of the

TABLE XX

Distribution of the Monetary Stock of the United States among the Public, the Banks, and the Federal Treasury on or about July 1, 1890 to 1911 *

	Actual Amounts in Millions of Dollars				Proportions of the Total Stock Held by the		
Year	Total Stock	In the Treasury as Assets	Outside of the Treasury	Held by Banks	Held by the Public	Treasury	Banks	Public
1890..	1,629	256	1,373	492	881	15.7%	31.0%	54.1%
1891..	1,623	180	1,443	504	931	11.1	31.0	57.9
1892..	1,689	151	1,538	585	953	9.0	34.6	56.4
1893..	1,673	142	1,531	517	1,014	8.5	30.9	60.6
1894..	1,726	144	1,582	680	902	8.3	39.4	52.3
1895..	1,726	217	1,509	620	889	12.6	35.9	51.5
1896..	1,703	294	1,414	556	858	17.2	32.6	50.2
1897..	1,805	266	1,539	649	890	14.7	36.0	49.3
1898..	1,959	236	1,723	713	1,010	12.0	36.4	51.6
1899..	2,076	286	1,790	748	1,042	13.8	36.0	50.2
1900..	2,223	285	1,938	787	1,151	12.8	35.4	51.8
1901..	2,362	308	2,054	828	1,226	13.0	35.1	51.9
1902..	2,439	314	2,125	875	1,250	12.9	35.9	51.2
1903..	2,560	317	2,243	881	1,362	12.4	34.4	53.2
1904..	2,674	284	2,390	1,016	1,374	10.6	38.0	51.4
1905..	2,754	295	2,459	1,024	1,435	10.7	37.2	52.1
1906..	2,937	333	2,604	1,043	1,561	11.3	35.5	53.2
1907..	2,115	343	2,772	1,139	1,633	11.0	36.6	52.4
1908..	3,378	341	3,037	1,394	1,643	10.1	41.3	48.6
1909 †	3,406	309	3,097	1,466	1,631	9.1	43.0	47.9
1910..	3,420	317	3,103	1,445	1,658	9.3	42.2	48.5
1911..	3,556	342	3,214	1,573	1,641	9.6	44.2	46.1

* Source: W. C. Mitchell, *Business Cycles, Memoirs of the University of California*, vol. 3, Berkeley, California, 1913, p. 296.
† Figures for April 30, instead of July 1st.

monetary stock that the proportion held by the public increased during both of these periods.

The movements of gold into and out of the Treasury also affected the amount of currency available for use by the public. It was pointed out by Andrew that the operations of the Treasury tended to lock up more gold in periods of prosperity, since taxes, especially customs duties, were then more prolific; and the opposite was true in depression.[1] Under Secretary Gage the deposits by the Treasury in banks was greatly increased, and the idle balances of the Treasury reduced.[2] But with Secretary Shaw a more decided policy came into effect. He aided the banks by transfers of gold to them from the Treasury during 1903, thus adding to the gold available for reserves.[3] During 1906 he aided the import of gold by permitting certain banks to anticipate arrivals by withdrawing cash immediately from the Treasury.[4] In effect this raised the gold import point by removing the interest cost. During the crisis of 1907 the Treasury supplied the New York banks with much of the cash sent to the interior.[5] Thus the evils indicated by Andrew were mitigated to a certain extent. But the Treasury operations introduced uncertainty as to the course which would be followed by the Secretary, and discouraged prudent preparation by bankers.[6]

In the panic period of 1907 the cash in circulation was supplemented by temporary media of various kinds.[7] The demands of the United States during prosperity and crisis on other countries for gold were consequently less than would otherwise have been the case. On the other hand, much of the cash removed from the banks was not utilized, but was hoarded by people.

During periods of prosperity the cash reserves of New York banks tended to fall, and they were replenished during depressions.[8] Thus these banks took up some of the slack in flows of gold from abroad to the interior circulation and bank reserves of the United States. Money rates, on the other hand, rose during expansion, and fell off in the succeeding depression.[9] The rise of money rates was certainly the proximate cause of the foreign influx of gold in 1906. Evidence to this effect is afforded by the large issue of finance bills drawn on London in that year.[10] In 1907 the major cause was the payment for agricultural exports.[11]

The studies of Professors Young and Mitchell reveal important dif-

[1] A. P. Andrew, "The Treasury and the Banks under Secretary Shaw," *Quarterly Journal of Economics*, vol. XXI, no. 4 (August, 1907), pp. 522–523.
[2] *Ibid.*, pp. 528–529.
[3] *Ibid.*, p. 540.
[4] *Ibid.*, pp. 543–547.
[5] See Sprague, *op. cit.*, p. 264, and Mitchell, *op. cit.*, p. 519.
[6] Andrew, *op. cit.*, pp. 560–562.
[7] See Mitchell, *op. cit.*, p. 518.
[8] See *ibid.*, pp. 332 and 347–351; A. H. Hansen, *Cycles of Prosperity and Depression in the United States, Great Britain and Germany, University of Wisconsin Studies in the Social Sciences and History*, No. 5, Madison, Wisconsin, 1921, p. 32; and Young, *op. cit.*, pp. 24–25.
[9] Hansen, *op. cit.*, p. 32, and Young, *op. cit.*, p. 24.
[10] See above, p. 143.
[11] See above, p. 145.

ferences in the cyclical movements of bank deposits in New York and in the country outside.[1] In periods of depression New York bank deposits and loans tended to increase, while in prosperity they declined. Banks outside of New York showed the opposite characteristics. The causes for this difference were the movement of idle cash in "other" banks to New York in depression and the withdrawal in expansion periods, and the shifting of bank deposits to and from "other" banks to New York.[2] There was also a marked cyclical variation in the volume of investments made by both the New York banks and those outside of New York.[3] In periods of depression the holding of securities increased, since surplus funds were large. But banks disposed of them ". . . when the demand for loans and discounts increased."[4]

The treatment of American data has been very brief, but several points stand out clearly. There was a tendency for currency to move into circulation during prosperity, and to reflow into the banks in depression. It is noticeable that the reflux occurred very quickly after the outbreak of the crisis, both in 1903-1904 and 1907-1908. Associated with this internal movement was a foreign flow of gold into the United States immediately before and during the crisis period, and a reflow from the country after the crises abated. Due to the presence of hoarded gold in the Treasury, and the large annual output from mines of the country, foreign flows were of less significance than in the case of England. Unless cyclical fluctuations of business were great, the repercussions upon international gold flows were not easily seen.

[1] See Young, *op. cit.*, pp. 28–29, and Mitchell, *op. cit.*, pp. 324–370.

[2] Young, *op. cit.*, p. 28. Mitchell remarks on the reasons for the sending of local bank funds to New York: "Perhaps the desire of the country banks to maintain their relatively high rates of interest upon local loans is a more potent factor than the unwillingness of their customers to borrow." *Op. cit.*, p. 362. See also Watkins, *op. cit.*, Chs. VIII and X, esp. pp. 322–345.

[3] Young, *op. cit.*, pp. 30–32.

[4] *Ibid.*, p. 30.

APPENDIX B

STATISTICAL APPENDIX

TABLE XXI

DEPOSIT AND CURRENT ACCOUNTS AND "CASH" RESERVES, NATIONAL
PROVINCIAL BANK OF ENGLAND, LIMITED, AND LONDON AND
WESTMINSTER BANK, LIMITED, MONTHLY, 1891–1913 *

(In thousands of pounds)

| | Deposit and Current Accounts † | | Cash on Hand and at Bank of England† | |
Approximate Date	National Provincial Bank of England, Limited	London and Westminster Bank, Limited	National Provincial Bank of England, Limited	London and Westminster Bank, Limited
Aug. 31, 1891.....	42,476	25,872	5,191	4,547
Sept. 30..........	42,026	24,634	4,601	3,821
Oct. 31..........	41,267	23,894	4,460	3,695
Nov. 30..........	40,481	...	4,419	4,335
Dec. 31..........	40,822	26,064	3,950	4,408
Jan. 31, 1892.....	40,727	24,475	5,182	3,999
Feb. 28..........	40,079	23,781	4,855	3,819
Mar. 31..........	39,987	22,976	5,027	3,679
Apr. 30..........	40,602	23,293	5,138	3,974
May 31..........	40,629	23,785	4,858	3,853
June 30..........	42,119	25,521	5,435	4,412
July 31..........	41,245	...	4,839	..
Aug. 31..........	41,766	25,165	5,319	4,571
Sept. 30..........	41,990	24,804	4,819	4,373
Oct. 31..........	42,341	24,325	5,194	3,987
Nov. 30..........	41,753	23,601	4,615	3,819
Dec. 31..........	41,805	24,676	4,907	3,953
Jan. 31, 1893.....	41,661	24,750	5,262	4,149
Feb. 28..........	41,461	24,292	4,966	3,960
Mar. 31..........	41,173	22,723	4,792	3,634
Apr. 30..........	41,407	...	5,254	..
May 31..........	41,959	21,022	5,636	3,723
June 30..........	42,570	22,397	5,205	3,863
July 31..........	42,596	23,791	4,929	4,875
Aug. 31..........	41,838	21,632	5,230	3,842
Sept. 30..........	41,615	21,867	5,035	3,685
Oct. 31..........	41,967	24,555	5,207	4,771
Nov. 30..........	41,889	25,089	5,054	4,107
Dec. 31..........	41,827	26,487	5,080	4,417

* SOURCE: *Economist.*
† The figures for the London and Westminster Bank, Limited, are carried only to 1909.

TABLE XXI (Continued)

Approximate Date	Deposit and Current Accounts		Cash on Hand and at Bank of England	
	National Provincial Bank of England, Limited	London and Westminster Bank, Limited	National Provincial Bank of England, Limited	London and Westminster Bank, Limited
Jan. 31, 1894.....	41,578	22,897	4,944	3,809
Feb. 28..........	41,627	22,404	5,256	3,717
Mar. 31..........	40,940	21,864	4,902	3,727
Apr. 30..........	40,964	22,282	4,799	3,724
May 31..........	41,077	....	5,036	..
June 30..........	42,288	23,993	4,837	3,899
July 31..........	42,266	....	4,974	..
Aug. 31..........	42,415	....	5,342	..
Sept. 30..........	42,288	25,096	5,080	4,177
Oct. 31..........	42,398	24,359	4,678	4,296
Nov. 30..........	42,544	23,929	5,358	3,867
Dec. 31..........	42,743	25,722	5,466	4,334
Jan. 31, 1895.....	42,820	25,137	5,059	4,445
Feb. 28..........	42,414	23,648	4,883	4,214
Mar. 31..........	42,238	23,212	4,865	4,201
Apr. 30..........	42,981	23,946	4,961	4,674
May 31..........	42,980	24,767	5,400	4,482
June 30..........	44,124	25,514	4,421	4,067
July 31..........	44,483	26,115	4,720	4,447
Aug. 31..........	44,661	....	5,175	..
Sept. 30..........	44,923	....	5,243	..
Oct. 31..........	45,326	27,122	5,035	4,532
Nov. 30..........	44,769	....	5,223	..
Dec. 31..........	45,642	28,227	5,770	5,002
Jan. 31, 1896.....	45,890	27,180	5,488	4,206
Feb. 28..........	45,184	....	4,937	..
Mar. 31..........	44,985	....	5,136	..
Apr. 30..........	45,703	26,598	4,716	4,601
May 31..........	45,064	26,168	4,737	4,016
June 30..........	46,595	27,828	5,075	3,866
July 31..........	46,111	28,003	4,753	4,460
Aug. 31..........	46,216	26,833	5,340	4,036
Sept. 30..........	46,806	26,434	5,418	4,234
Oct. 31..........	46,508	25,088	5,165	3,919
Nov. 30..........	45,967	24,326	4,840	3,919
Dec. 31..........	46,855	25,002	5,749	3,818
Jan. 31, 1897.....	46,879	24,928	4,736	3,908
Feb. 28..........	46,774	23,996	4,880	3,624
Mar. 31..........	46,886	23,937	4,970	3,682
Apr. 30..........	47,246	25,305	5,563	4,363
May 31..........	47,387	....	5,035	..
June 30..........	48,313	26,074	5,254	4,119
July 31..........	47,959	26,506	5,187	4,388
Aug. 31..........	47,478	26,466	5,441	4,354

TABLE XXI (*Continued*)

Approximate Date	Deposit and Current Accounts		Cash on Hand and at Bank of England	
	National Provincial Bank of England, Limited	London and Westminster Bank, Limited	National Provincial Bank of England, Limited	London and Westminster Bank, Limited
Sept. 30.........	47,259	26,208	5,396	4,261
Oct. 31.........	47,099	....	5,304	..
Nov. 30.........	46,761	24,195	5,301	4,179
Dec. 31.........	48,811	24,661	6,459	4,119
Jan. 31, 1898.....	48,920	24,555	5,562	3,625
Feb. 28.........	47,450	24,631	5,379	3,616
Mar. 31.........	47,342	23,991	5,630	4,214
Apr. 30.........	46,611	24,403	5,433	3,981
May 31.........	46,469	24,270	5,355	3,807
June 30.........	47,984	26,004	5,812	4,463
July 31.........	48,170	25,688	5,425	4,179
Aug. 31.........	47,742	26,694	5,464	4,438
Sept. 30.........	47,478	25,822	5,337	3,992
Oct. 31.........	47,372	25,378	5,312	3,818
Nov. 30.........	47,847	25,201	5,846	3,765
Dec. 31.........	49,286	27,578	6,196	4,872
Jan. 31, 1899.....	49,714	27,909	6,033	4,381
Feb. 28.........	49,150	28,527	5,555	4,179
Mar. 31.........	49,446	....	5,584	..
Apr. 30.........	49,485	....	5,959	..
May 31.........	50,538	....	5,951	..
June 30.........	51,293	27,886	6,010	4,820
July 31.........	51,095	25,734	5,946	4,027
Aug. 31.........	50,781	25,967	6,384	4,173
Sept. 30.........	50,043	25,754	6,626	4,454
Oct. 31.........	50,490	26,326	6,456	4,209
Nov. 30.........	49,583	25,702	6,377	4,081
Dec. 31.........	49,964	25,812	6,349	4,474
Jan. 31, 1900.....	49,789	25,563	6,600	3,917
Feb. 28.........	49,429	25,026	6,602	3,962
Mar. 31.........	48,984	24,893	6,457	4,090
Apr. 30.........	48,769	24,941	6,559	3,827
May 31.........	....	24,820	..	3,831
June 30.........	50,499	26,860	6,724	4,353
July 31.........	49,954	25,411	6,494	4,164
Aug. 31.........	49,714	25,026	6,922	3,747
Sept. 30.........	49,852	24,654	6,840	3,997
Oct. 31.........	50,186	25,358	7,097	3,966
Nov. 30.........	50,172	24,759	6,855	4,090
Dec. 31.........	51,084	26,278	7,398	4,203
Jan. 31, 1901.....	51,167	25,506	7,271	4,146
Feb. 28.........	49,943	25,148	7,021	3,946
Mar. 31.........	49,469	25,070	6,987	3,774
Apr. 30.........	49,006	25,320	7,021	4,290

TABLE XXI (*Continued*)

Approximate Date	Deposit and Current Accounts		Cash on Hand and at Bank of England	
	National Provincial Bank of England, Limited	London and Westminster Bank, Limited	National Provincial Bank of England, Limited	London and Westminster Bank, Limited
May 31..........	48,754	24,532	7,054	3,985
June 30..........	50,238	26,235	7,098	4,229
July 31..........	49,765	26,564	6,875	4,146
Aug. 31..........	49,378	25,017	7,136	3,734
Sept. 30..........	49,071	27,133	7,312	4,450
Oct. 31..........	49,423	27,242	7,442	4,241
Nov. 30...........	....	27,117	..	4,314
Dec. 31..........	50,640	27,155	7,561	4,788
Jan. 31, 1902.....	50,988	28,254	7,404	4,719
Feb. 28..........	51,436	27,282	7,675	4,546
Mar. 31..........	51,356	26,369	7,219	4,081
Apr. 30..........	50,973	27,439	7,327	4,238
May 31..........	49,820	26,349	7,719	3,663
June 30..........	51,154	27,372	8,175	4,743
July 31..........	51,376	27,315	7,490	4,308
Aug. 31..........	51,184	25,866	7,523	4,063
Sept. 30..........	51,468	25,243	7,436	3,770
Oct. 31..........	50,813	25,244	7,534	4,248
Nov. 30..........	50,438	24,898	7,638	4,098
Dec. 31..........	50,949	24,952	7,407	4,392
Jan. 31, 1903.....	51,599	25,490	7,841	3,849
Feb. 28..........	50,580	23,683	7,688	3,228
Mar. 31...........	49,785	22,549	7,432	3,411
Apr. 30..........	49,595	23,288	7,793	3,886
May 31..........	49,611	25,448	7,605	4,089
June 30..........	50,893	26,076	7,548	4,010
July 31..........	....	27,150	..	3,997
Aug. 31..........	51,126	....	7,837	..
Sept. 30..........	51,534	22,459	7,717	3,021
Oct. 3...........	....	22,823	..	3,016
Nov. 30..........	50,662	24,133	7,648	3,442
Dec. 31..........	50,360	27,990	7,491	4,657
Jan. 21, 1904.....	....	....	..	..
Feb. 28..........	50,598	24,675	7,681	3,701
Mar. 31..........	48,663	24,370	7,090	3,616
Apr. 30..........	48,834	24,274	7,234	3,916
May 31..........	49,055	24,138	7,485	3,951
June 30..........	50,511	24,425	7,284	3,481
July 31..........	50,618	24,686	7,386	3,423
Aug. 31..........	49,786	26,270	7,636	3,671
Sept. 30..........	50,068	26,805	7,634	3,967
Oct. 31..........	50,818	27,434	7,783	3,919
Nov. 30..........	....	....	..	..
Dec. 31..........	50,693	27,155	7,322	3,821

TABLE XXI (*Continued*)

Approximate Date	Deposit and Current Accounts		Cash on Hand and at Bank of England	
	National Provincial Bank of England, Limited	London and Westminster Bank, Limited	National Provincial Bank of England, Limited	London and Westminster Bank, Limited
Jan. 31, 1905	51,835	25,969	7,676	3,555
Feb. 28	50,980	25,837	7,741	3,896
Mar. 31	50,501	26,434	7,629	4,270
Apr. 30	50,673	26,470	7,133	3,878
May 31	50,533	25,133	7,685	3,560
June 30	52,345	27,817	7,797	4,322
July 31	53,017	28,156	7,601	4,260
Aug. 31	52,283	....	7,704	..
Sept. 30	52,761	27,072	8,134	3,657
Oct. 31	52,851	25,543	7,706	3,788
Nov. 30	51,358	23,863	7,828	3,454
Dec. 31	52,594	25,442	8,205	3,642
Jan. 31, 1906	52,987	24,004	8,004	3,387
Feb. 28	53,125	24,145	7,912	3,600
Mar. 31	52,356	24,571	9,007	3,137
Apr. 30	51,985	24,128	7,964	3,657
May 31	52,299	24,981	7,672	4,551
June 30	53,296	26,638	7,902	3,630
July 31	52,887	24,741	7,914	3,272
Aug. 31	53,026	26,680	7,955	4,107
Sept. 30	53,373	26,884	7,739	3,541
Oct. 31	53,522	26,382	7,857	3,781
Nov. 30	52,988	26,341	8,003	3,622
Dec. 31	54,242	28,474	8,603	3,829
Jan. 31, 1907	53,538	27,307	8,266	4,546
Feb. 28	54,694	27,817	8,087	4,672
Mar. 31	55,330	24,483	8,211	3,871
Apr. 30	54,684	26,203	8,240	4,261
May 31	54,484	26,439	8,114	4,001
June 30	54,850	....	7,809	3,781
July 31	55,733	....	8,550	..
Aug. 31	55,078	26,673	8,509	3,664
Sept. 30	54,211	26,712	8,248	3,353
Oct. 31	55,433	25,163	8,245	3,050
Nov. 30	53,888	25,850	8,227	3,195
Dec. 31	54,904	26,825	9,251	4,244
Jan. 31, 1908	55,953	....	8,885	..
Feb. 28	55,437	25,258	8,418	3,870
Mar. 31	54,833	26,133	8,302	4,018
Apr. 30	54,913	24,630	8,425	3,373
May 31	54,705	25,612	8,465	3,903
June 30	56,674	27,281	8,678	3,926
July 31	57,363	27,705	8,626	4,428
Aug. 31	57,042	27,414	8,305	4,152

TABLE XXI (*Continued*)

Approximate Date	Deposit and Current Accounts		Cash on Hand and at Bank of England	
	National Provincial Bank of England, Limited	London and Westminster Bank, Limited	National Provincial Bank of England, Limited	London and Westminster Bank, Limited
Sept. 30	58,096	26,782	9,147	3,258
Oct. 31	58,261	25,829	8,656	3,073
Nov. 30	58,577	23,216	9,199	2,883
Dec. 31	59,266	26,626	10,356	4,336
Jan. 31, 1909	58,746	23,833	9,547	3,687
Feb. 28	57,294	22,347	9,135	2,901
Mar. 31	56,200	22,816	8,414	3,345
Apr. 30	56,659	23,818	8,395	4,348
May 31	56,600	23,735	8,362	3,224
June 30	58,646	26,814	8,713	4,018
July 31	58,724	29,115	8,801	....
Aug. 31	59,112	....	8,807	....
Sept. 30	59,469	....	8,852	....
Oct. 31, 1909	58,625	....	8,520	....
Nov. 30	58,127	....	8,598	....
Dec. 31	59,542	....	10,073	....
Jan. 31, 1910	58,873	....	8,814	....
Feb. 28	60,824	....	8,987	....
Mar. 31	61,433	....	8,759	....
Apr. 30	62,240	....	9,323	....
May 31	62,720	....	9,493	....
June 30	62,387	....	9,850	....
July 31	63,877	....	9,678	....
Aug. 31	62,697	....	9,539	....
Sept. 30	63,625	....	9,577	....
Oct. 31	63,419	....	9,164	....
Nov. 30	63,410	....	9,296	....
Dec. 31	65,143	....	9,892	....
Jan. 31, 1911	65,355	....	9,532	....
Feb. 28	65,274	....	9,746	....
Mar. 31	63,744	....	9,742	....
Apr. 30	62,457	....	9,836	....
May 31	63,737	....	9,707	....
June 30	64,910	....	9,690	....
July 31	64,251	....	9,843	....
Aug. 31	63,026	....	9,720	....
Sept. 30	62,540	....	9,200	....
Oct. 31	62,212	....	9,333	....
Nov. 30	62,329	....	9,510	....
Dec. 31	62,892	....	9,676	....
Jan. 31, 1912	63,330	....	9,431	....
Feb. 28	62,056	....	9,567	....
Mar. 31	61,840	....	9,231	....
Apr. 30	62,207	....	9,489	....

TABLE XXI (*Continued*)

Approximate Date	Deposit and Current Accounts National Provincial Bank of England, Limited	Cash on Hand and at Bank of England National Provincial Bank of England, Limited
May 31	64,281	9,667
June 30	62,892	9,164
July 31	63,384	9,478
Aug. 31	64,058	9,511
Sept. 30	64,898	9,616
Oct. 31	64,476	9,465
Nov. 30	64,523	9,709
Dec. 31	65,660	10,292
Jan. 31, 1913	65,213	10,261
Feb. 28	64,704	9,992
Mar. 31	64,432	10,462
Apr. 30	63,324	9,863
May 31	64,143	10,115
June 30	66,606	9,964
July 31	65,704	10,349
Aug. 31	65,706	10,627
Sept. 30	65,655	10,127
Oct. 31	65,567	....
Nov. 30	65,668	....
Dec. 31	67,882	....

BIBLIOGRAPHY

BIBLIOGRAPHY

OFFICIAL DOCUMENTS

Great Britain, Annual Reports of the Deputy Master and Comptroller of the Mint, 1881-1916.
—— The Bank Charter Inquiry, 1832, on the Regulation of the Note Issues, Parliamentary Papers, 1831-1832, vol. VI.
—— Committee on Currency and Foreign Exchange After the War, First Interim Report. London, 1918.
—— Committee on Finance and Industry, Report. London, 1931.
—— Statistical Abstract for the United Kingdom, 1881-1913.
League of Nations, Gold Delegation of the Financial Committee, The Functioning of the Gold Standard, A Memorandum Submitted by Feliks Mlynarski. Geneva, 1931.
—— Interim Report. Geneva, 1930.
—— Report. Geneva, 1932.
—— Second Interim Report. Geneva, 1931.
—— Selected Documents. Geneva, 1930.
—— Selected Documents on the Distribution of Gold. Geneva, 1931.
United States, National Monetary Commission, 61st Congress, 2nd Session, The Bank of France in Its Relation to National and International Credit, by Maurice Patron, Sen. Doc. No. 494, Washington, 1910.
—— The Evolution of Credit and Banks in France, from the Founding of the Bank of France to the Present Time, by André Liesse, Sen. Doc. No. 522, Washington, 1909.
—— Financial Diagrams, compiled by A. P. Andrew, Sen. Doc. 509, Washington, 1910.
—— History of Crises Under the National Banking System, by O. M. W. Sprague, Sen. Doc. No. 538, Washington, 1910.
—— Interviews on the Banking and Currency Systems of England, Scotland, France, Germany, Switzerland, and Italy, Sen. Doc. No. 405, Washington, 1910.
—— Seasonal Variations in the Relative Demand for Money and Capital in the United States, by E. W. Kemmerer, Sen. Doc. No. 588, Washington, 1910.
—— Statistics for Great Britain, Germany, and France, 1867-1909, Sen. Doc. No. 578, Washington, 1910.
—— Statistics for the United States, 1867-1909, compiled by A. P. Andrew, Sen. Doc. No. 570, Washington, 1910.

BOOKS

Andreades, A. History of the Bank of England. London, 1909.
Angell, J. W. The Theory of International Prices. Cambridge, 1926.

Anonymous. The National Distress: Its Financial Origin and Remedy. London, 1848.
Bagehot, Walter. Lombard Street. London, 1870.
Beckhart, B. H. The Discount Policy of the Federal Reserve System. New York, 1924.
Bennison, W. The Causes of the Present Money Crisis Examined in Answer to the Pamphlet of Mr. J. Horseley Palmer. London, 1837.
Bonar, James (ed.). Letters of David Ricardo to Thomas Robert Malthus. Oxford, 1887.
Brown, W. A., Jr. England and the New Gold Standard, 1919–1926. New Haven, 1929.
Clare, George. A Money-Market Primer, and Key to the Exchanges. London, 1891.
Cobb, A. S. Banks' Cash Reserves: Threadneedle Street; A Reply to "Lombard Street." London, 1891.
Easton, H. T. Banks and Banking. London, 1896.
Einzig, Paul. International Gold Movements. London, 1929.
Elliot, A. D. The Life of George Joachim Goschen, First Viscount Goschen, 1831–1907. London, 1911.
Ely, R. T., Adams, T. S., Lorenz, M. O., and Young, A. A. Outlines of Economics. New York, 1908.
Escher, Franklin. Foreign Exchange Explained. New York, 1917.
Feavearyear, A. E. The Pound Sterling. Oxford, 1931.
Foxwell, H. S. Papers on Current Finance. London, 1919.
Fullarton, John. On the Regulation of Currencies. London, 1844.
Furniss, E. S. Foreign Exchange. Boston, 1922.
Giffin, Robert. Stock Exchange Securities; An Essay on the General Causes of Fluctuations in their Prices. London, 1879.
Gonner, E. C. K. (ed.). Economic Essays by David Ricardo. London, 1923.
Goschen, G. J. Essays and Addresses on Economic Questions. London, 1905.
—— The Theory of the Foreign Exchanges. London, 1861.
Gregory, T. E. The Gold Standard and Its Future. New York, 1932.
—— (ed.). Select Statutes & Reports Relating to British Banking, 1832–1928. London, 1929.
Hansen, A. H. Cycles of Prosperity and Depression in the United States, Great Britain and Germany, University of Wisconsin Studies in the Social Sciences and History, No. 5. Madison, 1921.
Hawtrey, R. G. The Art of Central Banking. London, 1932.
—— Currency and Credit. London, 1919.
Hobson, C. K. The Export of Capital. London, 1914.
Hulftegger, Otto. Die Bank von England. Zurich, 1915.
Jackson, F. H. To What Extent Has the Position of the Bank of England Changed in Recent Years in Comparison with — (a) Similar Institutions Abroad, and (b) London Clearing Bankers? London, 1903.
Jevons, W. S. Investigations in Currency and Finance. London, 1884.
Joplin, Thomas. An Examination of the Report of the Joint Stock Bank Committee. London, 1836.

Juglar, Clement. Du change et de la liberté d'emission. Paris, 1868.
—— Des crises commerciales et de leur retour périodique en France, en Angleterre, et aux États-Unis. Paris, 1862.
Keynes, J. M. Indian Currency and Finance. London, 1913.
—— A Treatise on Money. New York, 1930.
—— (ed.). Official Papers by Alfred Marshall. London, 1926.
Koch, Frederick. Der London Goldverkehr: Eine Volkswirtschaftliche Studie. Stuttgart and Berlin, 1905.
Laughlin, J. L. The Principles of Money. New York, 1903.
de Laveleye, Émile. Le marché monétaire et ses crises. Paris, 1866.
Lavington, F. The English Capital Market. London, 1921.
Leaf, Walter. Banking. London, 1926.
Lehfeldt, R. A. Gold, Prices and the Witwatersrand. London, 1919.
Loyd, S. J. (Lord Overstone). Further Reflections on the State of the Currency and the Action of the Bank of England. London, 1837.
—— The Petition of the Merchants, Bankers, and Traders of London, against the Bank Charter Act. London, 1847.
—— Reflections Suggested by a Perusal of Mr. J. Horseley Palmer's Pamphlet on the Causes and Consequences of the Pressure on the Money Market. London, 1837.
Macleod, H. D. The Theory and Practice of Banking. London, 1855-1856.
Mill, J. S. Principles of Political Economy. London, 1848.
Mitchell, W. C. Business Cycles, Memoirs of the University of California, vol. 3, Berkeley, 1913.
—— Business Cycles, the Problem and its Setting. New York, 1927.
Mlynarski, Feliks. Gold and Central Banks. New York, 1929.
Ohlin, Bertil. Interregional and International Trade. Cambridge, 1933.
Palgrave, R. H. Inglis. Bank Rate and the Money Market in England, France, Germany, Holland, and Belgium, 1844-1900. London, 1903.
Palmer, J. H. The Causes and Consequences of the Pressure upon the Money-Market. London, 1837.
Pigou, A. C. Industrial Fluctuations. London, 1927.
Rae, George. The Country Banker — His Clients, Cares and Work. London, 1885.
Ricardo, David. The Principles of Political Economy and Taxation. London, 1817.
Riefler, W. W. Money Rates and Money Markets in the United States. New York, 1930.
Royal Institute of International Affairs. The International Gold Problem. London, 1931.
—— Monetary Policy and the Depression. London, 1933.
Salomans, David. A Defense of the Joint-Stock Banks. London, 1837.
Shaw, W. A. Currency, Credit and the Exchanges During the Great War and Since (1914-1926). London, 1927.
—— The Theory and Principles of Central Banking. London, 1930.
Shirras, G. F. Indian Finance and Banking. London, 1920.
Spalding, W. F. The London Money Market. New York, 1922.
Steele, F. E. Present Day Banking. London, 1909.

Straker, Frederick. The Money Market. London, 1904.
Sykes, Joseph. The Amalgamation Movement in English Banking, 1825–1924. London, 1926.
Taussig, F. W. International Trade. New York, 1928.
Thorp, W. L. Business Annals. New York, 1926.
Torrens, Robert. A Letter to the Right Honourable Lord Viscount Melbourne on the Causes of the Recent Derangement in the Money Market and on Bank Reform. London, 1837.
Viner, Jacob. Canada's Balance of International Indebtedness, 1900–1913. Cambridge, 1924.
Watkins, L. L. Bankers' Balances. Chicago, 1929.
Whitaker, A. C.. Foreign Exchange. New York, 1920.
White, H. D. The French International Accounts, 1880–1913. Cambridge, 1933.
Wilson, Roland. Capital Imports and the Terms of Trade. Melbourne, 1931.
Withers, Hartley. The Meaning of Money. London, 1909.
—— Stocks and Shares. New York, 1915.
Wright, Quincy (ed.). Gold and Monetary Stabilization (Lectures on the Harris Foundation, 1932). Chicago, 1932.
Young, A. A. An Analysis of Bank Statistics for the United States. Cambridge, 1928.

PRINCIPAL PERIODICALS

The Bankers' Magazine, London.
Economic Journal.
The Economist.
Journal of the Institute of Bankers, London.
Quarterly Journal of Economics.

PERIODICAL ARTICLES

Ackland, Joseph. English Financial Panics — Their Causes and Treatment. Bankers' Magazine (August, 1896).
Andrew, A. P. The Influence of the Crops Upon Business in America. Quarterly Journal of Economics, vol. XX (May, 1906).
—— The Treasury and the Banks Under Secretary Shaw. Quarterly Journal of Economics, vol. XXI (August, 1907).
Angell, J. W. Equilibrium in International Trade, The United States, 1919–1926. Quarterly Journal of Economics, vol. XLII (May, 1928)
—— Review of Viner's Canada's Balance of International Indebtedness, 1900–1913. Political Science Quarterly, vol. XL (June, 1925).
Atchison, W. J. On the Ratio a Banker's Cash Reserve Should Bear to his Liability on Current and Deposit Account. Journal of the Institute of Bankers, vol. VI (June, 1885).
Bullock, C. J., Williams, J. H., and Tucker, R. S. The Balance of Trade of the United States. The Review of Economic Statistics (July, 1919).
Hansard, Luke. The Efficiency of Reserves. Bankers' Magazine (March, 1901).

Hawtrey, R. G. London and the Trade Cycle. American Economic Review, vol. xix, Supplement (March, 1929).

Macrosty, H. W. Submerged Information, Journal of the Royal Statistical Society, vol. xc (New Series), (1927).

Martin, J. B. The Evolution of our Banking System, Economic Journal, vol. 1 (September, 1891).

Persons, W. M. Cyclical Fluctuations of the Ratio of Bank Loans to Deposits, 1867–1914. Review of Economic Statistics, vol. vi (October, 1924).

Persons, W. M., Silberling, N. J., and Berridge, W. A. An Index of British Economic Conditions, 1903–1914. Review of Economic Statistics, Preliminary vol. iv, Supplement no. 2 (June, 1922).

Silverman, A. G. Some International Trade Factors for Great Britain. Review of Economic Statistics, vol. xiii (August, 1931).

Sprague, O. M. W. The Distribution of Money Between the Banks and the People Since 1893. Quarterly Journal of Economics, vol. xviii (August, 1904).

Thomas, D. S. An Index of British Business Cycles. Journal of the American Statistical Association, New Series, vol. xxi (March, 1926).

Tritton, J. H. The Short Loan Fund of the London Money Market. Journal of the Institute of Bankers, vol. xxiii (March, 1902).

Whitaker, A. C. The Ricardian Theory of Gold Movements and Professor Laughlin's Views of Money. Quarterly Journal of Economics, vol. xviii (February, 1904).

INDEX

INDEX

Internal gold movements. *See* Currency.

International gold movements, classification of, 24 ff.; classical theory of, 3 ff., Ch. II, *passim*, 154 ff., Ch. IX, *passim*; neo-classical theory of, 3 ff., 9 ff.; relation of to discount rates. *See* Contents, Chs. II and IX; relation of to capital movements, 9 ff., 16. *See also* Mechanism of adjustment, and Gold movements.

Jackson, F. H., 53 n., 88 n.
Jevons, W. S., 60 n.
Joint stock banks, balances at Bank of England of, 80 ff., 118 ff., 129, 133 n.; cash in hand of, 82 ff., 133 n.; co-operation of with Bank of England, 119, 142, 150, 158 ff.; deposits of, 91 ff.; effect of amalgamations on reserves and deposits of, 87, 94; gold holdings of, 88 ff., 130, 133 n., 149, 168; reserves of, 80 ff., 99, 118 ff., 129, 133 n., 168.
Joplin, T., 32 n., 34 n.
Juglar, C., 35 ff.

Kemmerer, E. W., 185 ff.
Keynes, J. M., 6 n., 19 n., 55 n., 82 n., 86 n., 124 n., 180 n., 182.

Laughlin, J. L., 19, 21 ff.
de Laveleye, É., 35 ff.
Lavington, F., 50 n.
Leaf, W., 136 n.
Lehfeldt, R. A., 55 n.
Liesse, A., 144 n., 146 n.
London and Midland Bank, 89 ff.
London and Westminster Bank, 94 ff., 197 ff.
London Customs House, statistics of gold imports and exports of, 45 ff.
London money market. *See* Contents, Chs. V, VI, VII, and VIII.
Loyd, S. J. (Lord Overstone), 20 n., 31 ff., 34, 35.

Macleod, H. D., 19 n.
Macrosty, H. W., 81 n., 100 n.
Market discount rates. *See* Discount rates.
Marshall, A., 36 ff., 152 n., 167 n., 168 n.

Martin, J. B., 165.
Mechanism of adjustment, classical theory of, 3 ff., Ch. II, *passim*, 154 ff., Ch. IX, *passim*; neo-classical theory of, 3 ff., 9 ff.; part played by shifts of demand schedules, 4, 30 ff., 171 ff.; by transfers of purchasing power, 4, 26 ff.; relation of banking systems to, 3 ff., Ch. II, *passim*, 77 ff., 91, 100, 121, 154 ff., 167 ff., Ch. IX, *passim*, 185. *See also* International gold movements.
Mill, J. S., 8.
Mitchell, W. C., 40 n., 70 n., 74, 135 n., 136 n., 140 n., 144 n., 169 n., 185 n., 186 n., 191, 192 n., 194 n., 195 ff.
Mlynarski, F., 19 n.

National Provincial Bank, 94 ff., 197 ff.

Ohlin, B., 4 n., 30 n.

Palgrave, R. H. Inglis-, 60 n., 74, 81 n., 94 n.
Palmer, J. H., 32 n.
Patron, M., 144 n., 146 n.
Persons, W. M., 40, 81, 186 n.
Pigou, A. C., 78 n., 94.
Prices, index of, for England, 40 ff.; relation to costs, Chs. I and II, *passim*, 171.
Price-specie flow analysis. *See* Mechanism of adjustment.
Purchasing power, transfers of, 4, 26 ff.

Rae, G., 161 n., 162 n., 166 n.
Ricardo, D., 7.
Riefler, W. W., 181.

Salomons, D., 32 n.
Sectional price levels, 9 ff., 15 ff., 30 ff.
Shaw, W. A., 55 n.
Shirras, G. F., 55 n.
Short-term borrowing. *See* Foreign balances.
Silberling, N. J., 40, 81.
Silver coin, in England, 50 ff., 73 ff.
Silverman, A. G., 10 n., 40 n., 47.
Spalding, W. F., 54 n., 158 n., 162.
Sprague, O. M. W., 143 n., 145 n., 186 n., 193 n., 195 n.
Steele, F. E., 159 n.